AF539329

Plant Cytogenetic
Breeding and Evolution

Plant Cytogenetic
Breeding and Evolution

Sanjay Sharma

RANDOM PUBLICATIONS
NEW DELHI (INDIA)

Plant Cytogenetic: Breeding and Evolution

ISBN 978-93-5111-133-7

Published in 2013 in India by

RANDOM PUBLICATIONS

4376-A/4B, Gali Murari Lal, Ansari Road
New Delhi-110 002
Phone : +9111-43580356, 011-23289044
e-mail: randompublications@hotmail.com

Type Setting by : Friends Media, Delhi-110089
Printed at : Thomson Press (India) Ltd.

Preface

Plant Cytogenetics, Breeding and Evolution plant cytogenetics comprises a topic of broad interest and increasing importance in plant science.

Major genome research programmes are underway to isolate expressed genes of corn and to DNA sequence at least the gene-rich regions. We are developing a highly efficient method for mapping DNA sequences to chromosome and sub-chromosomal regions based on lines derived from crosses between oats and corn. Partial hybrids have been obtained that contain an entire set of oat chromosomes along with one corn chromosome; lines are available that have each chromosome of corn individually added to the oat genome.

Plant breeding is the art and science of changing the genetics of plants for the benefit of mankind. Plant breeding can be accomplished through many different techniques ranging from simply selecting plants with desirable characteristics for propagation, to more complex molecular techniques.

Plant breeding has been practiced for thousands of years, since near the beginning of human civilization. It is now practiced worldwide by individuals such as gardeners and farmers, or by professional plant breeders employed by organizations such as government institutions, universities, crop-specific industry associations or research centers.

International development agencies believe that breeding new crops is important for ensuring food security by developing new varieties that are higher-yielding, resistant to pests and diseases, drought-resistant or regionally adapted to different environments and growing conditions.

The book provides a unique combination of historical and modern subject matter, revealing the central role of plant cytogenetics in plant genetics and genomics as currently practiced.

—Author

Contents

1

Breeding Systems

PLANT BREEDING

One plant may be more favourably situated, getting more light, more manure or water, and because of this is bigger, leafier, or more productive. Another, less well favoured, suffering from paucity of supplies, imbalance of nutrients, or disease, is smaller or deformed or discoloured.

The science of genetics, or heredity, is concerned with questions involved in the transmission, from one generation to the next, of all the attributes which gives each its individuality. In short, genetics seeks to explain why sometimes "like begets like," but at other times "begets the unlike." There are two classes of factors which condition the appearance and behaviour of the individual. On the one hand, there are the factors of the environment.

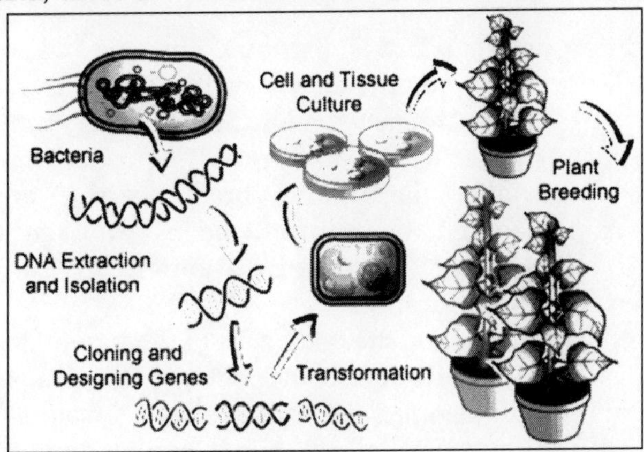

Fig. Plant Breeding

The two plants, had they been grown under similar circumstances, might have appeared and behaved alike. On the other hand, there are differences due to inheritance. There are tall varieties and dwarf varieties of peas; wheat varieties with either white grain or red grain; clover with hairy leaves or non-hairy leaves. The production of these characters is not due to factors of the environment, but to factors inside the plant which were transmitted to it from its parents.

The characters of the plant, then, as we see them are the result of an interaction between the factors of environment and the internal, hereditable, or genetical factors. The action of the external factors is' temporary, and produces only *fluctuations* in the course of development; the internal factors are passed on to the next generation unaltered.

The genetical factors are quasi-permanent units passing from generation to generation, and conditioning in each, definite *variations* between individuals differently endowed. Where genetical questions are involved comparisons between plants should be made only between individuals which have been treated alike, and therefore where variation and not fluctuation has been the cause of any differences between them.

Another point which must be clarified is the difference between propagation and reproduction. Propagation is the re-formation of a whole plant from a portion of the body of the "parent." Only body (somatic) cells are involved; only mitotic or somatic cell divisions occur; there is no fusion of nuclei. Hence, all the cells of the "daughter" plants are constituted or endowed exactly as the cells of the "mother," and contain the same set of factors.

No matter whether the propagation is natural, as by runners, stolons, bulbs, corms, etc., or artificially produced by cuttings, grafts, buds, it involves only the splitting up of one body into parts and the establishment of these parts as separate entities.

PROPAGATION

By propagation, *one individual* is distributed in time and space. An apple variety, "James Grieve," is bred only once. One tree is grown from a true seed. Every unit of "James Grieve"in the world is merely a part of the original tree, or a part of a part of it grafted on to appropriate roots. Similarly, with a commercial variety of potato, the variety is bred once, then tubers (branches of the body) are distributed. Units produced by propagation are called propagants, or clones. They are identical genetically, and, subject to fluctuation, identical in appearance.

Conversely, in reproduction the body cells of the parent do not enter into the process directly; meiotic division comes between the somatic cell and the gamete, which in its chromosomes carries the genetical factors. The gamete is not constituted like the parental cell, though it derives its constitution from it. Further, in reproduction there is fusion of two gametes at fertilization to give the zygote from which the body of the new generation derives.

The two gametes which fuse may have come each from parents of quite different genetical constitution, and so the offspring of the mating may not wholly resemble either parent but each in part. Excluding the question of mutation which will be discussed later, propagation gives complete uniformity in the progeny; reproduction holds out the chance of variation.

CROSSING

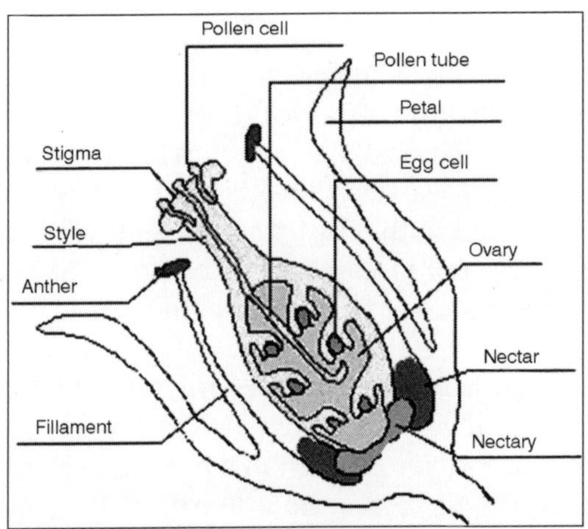

Fig. Pollen Germinates

The genetical factors which pass from generation to generation must be carried in the gametes, the particular nuclei seen in the pollen grain and embryo sac. To effect a cross between two plants, suipollen must be transferred from the stamens of one to the stigma of the other, precautions being taken that no other pollen is present. All being well, the pollen germinates and fertilization follows.

The two plants so "crossed"constitute the "first parental generation," or Pi, and their immediate progeny, the "first filial generation," or Fi. If the pollen supply used in fertilizing the Fi is restricted to its own spores, and it becomes self-fertilized or "selfed," the next generation is known as Fs.

Simple Cross

By appropriate crossing and selfing the behaviour in heredity of the different factors may be followed. A very simple case is seen in wheat. Some varieties have the spikelets arranged on the rachis, with an interval between. In others, the spikelets are placed quite close together and the "head"is compact. The one is said to have a "lax head," the other a "dense head." If two plants, one from each of these types, is crossed, the Fi is intermediate as regards this character. When the Fi is selfed, the F2 shows three classes of plant.

In one class all the plants have lax heads; in the second they are all dense; while in the third they resemble the Fi, and are intermediate in character. The cross is made either way, the dense head type or the lax head being used as the female, it matters not; if sufficient plants are grown in Fs, the numbers in the three classes will always bear the same numerical relationship to each other: 25 per cent, lax, 50 per cent, intermediate, 25 per cent, dense; or the ratio 1:2:1.

DOMINANCE

When other characters are considered in a hybridizing experiment, the Fi may not be intermediate but resemble one of the parents. There are bearded wheats and non-bearded. That is, the "chaff" which envelopes the grain may have a long tapering awn (bearded), or this may be absent (beardless). When a bearded wheat is crossed with a beardless one, the Fi is bearded. The gene for bearded dominates in the Fi. It is said to be dominant to the non-bearded gene, which conversely is said to be recessive. In Fs only two classes appear—beardless and bearded.

The numbers of plants in ttyese classes are in the ratio of 3: 1. In Plate 120 specimens from the actual plants in such a cross have been mounted in such a way as to illustrate the behaviour seen in a mating of this type. In the lowest line of plants (the Fa) it will be seen that there are twelve bearded plants to four without beards; that is, the numbers in the two classes are in the ratio of 3: 1. This is quite simply interpreted when it is remembered that one "dose"of bearded "swamps"or masks the presence of one dose of beardless.

It is clear that the 3: 1 ratio is merely the 1:2:1 with the first two classes amalgamated by the effect of dominance. Only one plant (the homozygote) in every three of the bearded class will breed true. In this particular case the heterozygote produces much smaller awns than those borne by the homozygote. This partial production of the character is called partial dominance. When the heterozygote is quite similar to the homozygote the dominance is said to be total.

Chromosome Gene

A plant shows hundreds of units of character and must therefore have hundreds of pairs of genes, yet it has only a limited number of chromosomes in each nucleus; in bread wheat there are twenty-one pairs in the soma or twenty-one individuals in the gamete. Many other plants have fewer chromosomes, yet these bodies have to carry all the genes. It is known that each chromosome carries a number of genes, and these are distributed up and down the chromosome in linear order like beads on a string.

Each chromosome always carries the same gene set. The distance between neighbour genes on a chromosome is constant, and the order in which they occur from one end to the other never varies in normal cases. Linkage would prohibit the possibility of new combinations being bred, were it not for the fact that at syndesis in meiosis, the chromosomes of homologous pairs become closely intertwined, and in separating give the appearance of having broken and united with an exchange of parts between the two members of a chromosome pair. The points where they cross-over each other in the intertwining, called a *chiasma,* can be seen under the microscope.

On the left is the original chromosome pair; next the two intertwined; next the two chromosomes are shown reformed with exchange of parts, while on the extreme right are the chromosomes as they go forward to complete mieosis and pass to the gametes.

Phenotype and Genotype

A plant described on the basis of its external appearance is called the phenotype. If described by reference to the genes inside it is called the genotype. In the Fs, then, there are bearded phenotypes, some of which (the heterozy-gotes) are genotypically bearded and beardless, while the others (homozygotes) are phenotypically and genotypically beardless.

Genetical Purity

In order to distinguish clearly between a homozygous and a heterozygous bearded, or, in other words, to distinguish between the two possible genotypes, the plants must be self-fertilized and bred-on. The progeny of the two genotypes will differ. On self-fertilization the homozygote produces only one class of plant, or, in other words, it "breeds true." On the other hand, the heterozygote segregates. This "proving"of the plant by reference to the character of its offspring is called the progeny test. Where only a small number of plants can be grown, it may so chance that following on self-fertilization the recessive form does not happen to be produced.

If, however, the suspected heterozygote, instead of being self-fertilized, is crossed with the pure recessive parental type, the chances of the recessive condition appearing in the progeny is increased. Every pollen grain from a beardless plant carries the gene beardless.

This, supplied to a heterozygous plant, ensures that every ovule carrying the recessive gene will produce a beardless individual. In short, the cross of a heterozygote to the recessive ensures a 50/50 distribution in the two classes. Crossing back to the recessive parental type is a special case of progeny testing called the *back-cross.*

The phenomenon of segregation and recombination of these genetical factors was discovered in the garden pea by Mendel. The ratios which he demonstrated are called Mendelian ratios, and the study of their various forms, Mendelism. Very many other but similar cases have been described in all sorts of organisms by many investigators.

Sometimes dominance appears, sometimes the Fi is intermediate. When the parents of a cross differ in two pairs of characters, both showing dominance, four classes of plant appear.

This is due to two 3:1ratios being superimposed. Re-examination of the photograph will show that the left-hand awnless parent (variety Iron) has also white chaff while the right-hand bearded parent has red chaff. These two parents differ in two characters—awnedness and chaff colour.

The Fi is seen to resemble the "April" parent in that it is both bearded and red-coloured in the chaff. When the Fi proceeds to form gametes the two pairs of genes involved segregate quite independently of each other, and therefore produce equal.

Quantities of pollen grains in four different classes, each class differing from the others in gene content. Similarly there are four classes of female gamete.

THE GENE

The mechanism which lies behind this segregation and recombination of factors is simple. It is now established that the genetical "factor"has a material basis in a piece of the structure of a chromosome. The name for this material basis of the theoretical factor is the gene. Each gene is known to occupy a definite place (the locus) in a definite chromosome. The chromosomes are paired bodies, and in a diploid (somatic) cell-nucleus there are two homologues, each carrying the members of the same gene pairs. The chromosomes of each homologous pair at meiosis part company.

This means that every gamete will receive one chromosome from each chromosome pair, and therefore one gene from each gene pair. At fertilization the chromosome pairs, and therefore the gene pairs, are reconstituted. The gene controlling the arrangement of spikelets on the rachis exists in two forms or potentialities. There is one potential for the lax condition, and the other potential for the dense condition. All the gametes of the lax parent will carry the gene for lax (L), and all the gametes of the dense parent will carry the gene for dense (D). These two different aspects of the gene pair meet at the fusion of the crossed gametes, and are therefore present in the somatic nuclei of the Fi, where they exert their influence.

When the Fi in its turn proceeds to form gametes the chromosome homologues part, and therefore the genes they bear part also. One half of the pollen will receive the gene aspect for lax, the other half for dense. Similarly, 50 per cent, of the ovules will carry "lax," and 50 per cent. "dense." On self-fertilization of the Fi, it depends entirely on chance whether a pollen grain carrying "lax"meets an ovule carrying "dense"or "lax." Similarly for pollen grains carrying "dense," as there are equal numbers of dense and lax carriers in both the male and female gametes.

The Heterozygote

These hybrid plants whose somatic nuclei develop from a zygote containing both aspects of the gene pair *(e.g.* lax-producing and dense-producing) are known as heterozygotes, and are said to show the heterozygous condition. On the other hand, plants whose nuclei have two "doses"of the one aspect of the pair are homozygous, and such plants are homozygotes. The homozy-gote breeds true when self-fertilized, for all its gametes are alike. The heterozygote when self-fertilized shows segregation.

When it forms pollen and ovules the two aspects of the gene pair part company and, depending how they meet for fertilization, produce plants of different character. The essential point here is that gene pairs segregate at reduction division because their chromosomes segregate. They recom-bine at fertilization because at that point in the life-cycle the chromosome pairs are reconstituted. This explains why there are three classes in F2, and the fact that when large numbers are grown in this second filial generation the numbers in each class occur with the bounds of reassortment on the basis of pure chance in the ratio of 1: 2: 1.

GENETICS OF PLANT BREEDING

Plant breeding is the art and science of changing the genetics of plants for the benefit of mankind. Plant breeding can be accomplished through many different techniques ranging from simply selecting plants with desirable characteristics for propagation, to more complex molecular techniques

Plant breeding has been practiced for thousands of years, since near the beginning of human civilization. It is now practiced worldwide by individuals such as gardeners and farmers, or by professional plant breeders employed by organizations such as government institutions, universities, crop-specific industry associations or research centers.

International development agencies believe that breeding new crops is important for ensuring food security by developing new varieties that are higher-yielding, resistant to pests and diseases, drought-resistant or regionally adapted to different environments and growing conditions.

DOMESTICATION

Plant breeding in certain situations may lead to the domestication of wild plants. Domestication of plants is an artificial selection process conducted by humans to produce plants that have more desirable traits than wild plants, and which renders them dependent on artificial (usually enhanced) environments for their continued existence. The practice is estimated to date back 9,000-11,000 years. Many crops in present day cultivation are the result of domestication in ancient times, about 5,000 years ago in the Old World and 3,000 years ago in the New World. In the Neolithic period, domestication took a minimum of 1,000 years and a maximum of 7,000 years. Today, all of our principal food crops come from domesticated varieties. Almost all the domesticated plants used today for food and agriculture were domesticated in the centers of origin. In these centers there is still a great diversity of closely related wild plants, so-called crop wild relatives, that can also be used for improving modern cultivars by plant breeding.

A plant whose origin or selection is due primarily to intentional human activity is called a cultigen, and a cultivated crop species that has evolved from wild populations due to selective pressures from traditional farmers is called a landrace. Landraces, which can be the result of natural forces or domestication, are plants (or animals) that are ideally suited to a particular region or environment. An example are the landraces of rice, *Oryza sativa* subspecies *indica*, which was developed in South Asia, and *Oryza sativa* subspecies *japonica*, which was developed in China.

CLASSICAL PLANT BREEDING

Classical plant breeding uses deliberate interbreeding (crossing) of closely or distantly related individuals to produce new crop varieties or lines with desirable properties. Plants are crossbred to introduce traits/genes from one variety or line into a new genetic background. For example, a mildew-resistant pea may be crossed with a high-yielding but susceptible pea, the goal of the cross being to introduce mildew resistance without losing the high-yield characteristics. Progeny from the cross would then be crossed with the high-yielding parent to ensure that the progeny were most like the high-yielding parent, (backcrossing). The progeny from that cross would then be tested for yield and mildew resistance and high-yielding resistant plants would be further developed. Plants may also be crossed with themselves to produce inbred varieties for breeding.

Classical breeding relies largely on homologous recombination between chromosomes to generate genetic diversity. The classical plant breeder may also makes use of a number of *in vitro* techniques such as protoplast fusion, embryo rescue or mutagenesis to generate diversity and produce hybrid plants that would not exist in nature.

Traits that breeders have tried to incorporate into crop plants in the last 100 years include:

1. Increased quality and yield of the crop
2. Increased tolerance of environmental pressures (salinity, extreme temperature, drought)
3. Resistance to viruses, fungi and bacteria
4. Increased tolerance to insect pests
5. Increased tolerance of herbicides

BEFORE WORLD WAR II

Intraspecific hybridization within a plant species was demonstrated by Charles Darwin and Gregor Mendel, and was further developed by geneticists and plant breeders. In the United Kingdom in the 1880s, it was the pioneering work of Gartons Agricultural Plant Breeders. In the early 20th century, plant breeders realized that Mendel's findings on the non-random nature of inheritance could be applied to seedling populations produced through deliberate pollinations to predict the frequencies of different types.

From 1904 to World War II in Italy Nazareno Strampelli created a number of wheat hybrids. His work allowed Italy to increase hugely crop production during the so called "Battle for Grain" (1925–1940) and some varieties was exported in foreign countries, as Argentina, Mexico, China and others. After the war, the work of Strampelli was quickly forgotten, but thanks to the hybrids he created, Norman Borlaug was able to move the very first steps of the Green Revolution.

In 1908, George Harrison Shull described heterosis, also known as hybrid vigour. Heterosis describes the tendency of the progeny of a specific cross to outperform both parents.

The detection of the usefulness of heterosis for plant breeding has led to the development of inbred lines that reveal a heterotic yield advantage when they are crossed. Maize was the first species where heterosis was widely used to produce hybrids.

By the 1920s, statistical methods were developed to analyze gene action and distinguish heritable variation from variation caused by environment. In 1933, another important breeding technique, cytoplasmic male sterility (CMS), developed in maize, was described by Marcus Morton Rhoades. CMS is a maternally inherited trait that makes the plant produce sterile pollen. This enables the production of hybrids without the need for labor intensive detasseling.

These early breeding techniques resulted in large yield increase in the United States in the early 20th century. Similar yield increases were not produced elsewhere until after World War II, the Green Revolution increased crop production in the developing world in the 1960s.

AFTER WORLD WAR II

Following World War II a number of techniques were developed that allowed plant breeders to hybridize distantly related species, and artificially induce genetic diversity.

When distantly related species are crossed, plant breeders make use of a number of plant tissue culture techniques to produce progeny from otherwise fruitless mating. Interspecific and intergeneric hybrids are produced from a cross of related species or genera that do not normally sexually reproduce with each other.

These crosses are referred to as *Wide crosses*. For example, the cereal triticale is a wheat and rye hybrid. The cells in the plants derived from the first generation created from the cross contained an uneven number of chromosomes and as result was sterile. The cell division inhibitor colchicine was used to double the number of chromosomes in the cell and thus allow the production of a fertile line.

Failure to produce a hybrid may be due to pre- or post-fertilization incompatibility. If fertilization is possible between two species or genera, the hybrid embryo may abort before maturation. If this does occur the embryo resulting from an interspecific or intergeneric cross can sometimes be rescued and cultured to produce a whole plant. Such a method is referred to as *Embryo Rescue*. This technique has been used to produce new rice for Africa, an interspecific cross of Asian rice *(Oryza sativa)* and African rice *(Oryza glaberrima)*.

Hybrids may also be produced by a technique called protoplast fusion. In this case protoplasts are fused, usually in an electric field. Viable recombinants can be regenerated in culture.

Chemical mutagens like EMS and DMS, radiation and transposons are used to generate mutants with desirable traits to be bred with other cultivars - a process known as *Mutation Breeding*. Classical plant breeders also generate genetic diversity within a species by exploiting a process called somaclonal variation, which occurs in plants produced from tissue culture, particularly plants derived from callus. Induced polyploidy, and the addition or removal of chromosomes using a technique called chromosome engineering may also be used.

When a desirable trait has been bred into a species, a number of crosses to the favoured parent are made to make the new plant as similar to the favoured parent as possible.

Returning to the example of the mildew resistant pea being crossed with a high-yielding but susceptible pea, to make the mildew resistant progeny of the cross most like the high-yielding parent, the progeny will be crossed back to that parent for several generations. This process removes most of the genetic contribution of the mildew resistant parent. Classical breeding is therefore a cyclical process.

With classical breeding techniques, the breeder does not know exactly what genes have been introduced to the new cultivars. Some scientists therefore argue that plants produced by classical breeding methods should undergo the same safety testing regime as genetically modified plants. There have been instances where plants bred using classical techniques have been unsuitable for human consumption, for example the poison solanine was unintentionally increased to unacceptable levels in certain varieties of potato through plant breeding. New potato varieties are often screened for solanine levels before reaching the marketplace.

MODERN PLANT BREEDING

Modern plant breeding uses techniques of molecular biology to select, or in the case of genetic modification, to insert, desirable traits into plants.

STEPS OF PLANT BREEDING

The following are the major activities of plant breeding;

1. Creation of variation
2. Selection
3. Evaluation
4. Release
5. Multiplication
6. Distribution of the new variety

MARKER ASSISTED SELECTION

Sometimes many different genes can influence a desirable trait in plant breeding. The use of tools such as molecular markers or DNA fingerprinting can map thousands of genes. This allows plant breeders to screen large populations of plants for those that possess the trait of interest. The screening is based on the presence or absence of a certain gene as determined by laboratory procedures, rather than on the visual identification of the expressed trait in the plant.

REVERSE BREEDING AND DOUBLED HAPLOIDS (DH)

A method for efficiently producing homozygous plants from a heterozygous starting plant, which has all desirable traits. This starting plant is induced to produce doubled haploid from haploid cells, and later on creating homozygous/doubled haploid plants from those cells. While in natural offspring genetic recombination occurs and traits can be unlinked from each other, in doubled haploid cells and in the resulting DH plants recombination is no longer an issue. There, a recombination between two corresponding chromosomes does not lead to un-linkage of alleles or traits, since it just leads to recombination with its identical copy. Thus, traits on one chromosome stay linked. Selecting those offspring having the desired set of chromosomes and crossing them will result in a final F1 hybrid plant, having exactly the same set of chromosomes, genes and traits as the starting hybrid plant. The homozygous parental lines can reconstitute the original heterozygous plant by crossing, if desired even in a large quantity. An individual heterozygous plant can be converted into a heterozygous variety (F1 hybrid) without the necessity of vegetative propagation but as the result of the cross of two homozygous/doubled haploid lines derived from the originally selected plant. patent

GENETIC MODIFICATION

Genetic modification of plants is achieved by adding a specific gene or genes to a plant, or by knocking down a gene with RNAi, to produce a desirable phenotype. The plants resulting from adding a gene are often referred to as transgenic plants. If for genetic modification genes of the species or of a crossable plant are used under control of their native promoter, then they are called cisgenic plants. Genetic modification can produce a plant with the desired trait or traits faster than classical breeding because the majority of the plant's genome is not altered. To genetically modify a plant, a genetic construct must be designed so that the gene to be added or removed will be expressed by the plant. To do this, a promoter to drive transcription and a termination sequence to stop transcription of the new gene, and the gene or genes of interest must be introduced to the plant. A marker for the selection of transformed plants is also included. In the laboratory, antibiotic resistance is a commonly used marker: Plants that have been successfully transformed will grow on media containing antibiotics; plants that have not been transformed will die. In some instances markers for selection are removed by backcrossing with the parent plant prior to commercial release.

The construct can be inserted in the plant genome by genetic recombination using the bacteria *Agrobacterium tumefaciens* or *A. rhizogenes*, or by direct methods like the gene gun or microinjection. Using plant viruses to insert genetic constructs into plants is also a possibility, but the technique is limited by the host range of the virus. For example, Cauliflower mosaic virus (CaMV) only infects cauliflower and related species. Another limitation of viral vectors is that the virus is not usually passed on the progeny, so every plant has to be inoculated.

The majority of commercially released transgenic plants are currently limited to plants that have introduced resistance to insect pests and herbicides. Insect resistance is achieved through incorporation of a gene from *Bacillus thuringiensis* (Bt) that encodes a protein that is toxic to some insects. For example, the cotton bollworm, a common cotton pest, feeds on Bt cotton it will ingest the toxin and die. Herbicides usually work by binding to certain plant enzymes and inhibiting their action. The enzymes that the herbicide inhibits are known as the herbicides *target site*. Herbicide resistance can be engineered into crops by expressing a version of *target site* protein that is not inhibited by the herbicide. This is the method used to produce glyphosate resistant crop plants. Genetic modification of plants that can produce pharmaceuticals (and industrial chemicals), sometimes called *pharmacrops*, is a rather radical new area of plant breeding.

PARTICIPATORY PLANT BREEDING

The development of agricultural science, with phenomenon like the Green Revolution arising, have left millions of farmers in developing countries, most of whom operate small farms under unstable and difficult growing conditions, in a precarious situation. The adoption of new plant varieties by this group has been hampered by the constraints of poverty and the international policies promoting an industrialized model of agriculture. Their response has been the creation of a novel and promising set of research methods collectively known as participatory plant breeding. Participatory means that farmers are more involved in the breeding process and breeding goals are defined by farmers instead of international seed companies with their large-scale breeding programmes. Farmers' groups and NGOs, for example, may wish to affirm local people's rights over genetic resources, produce seeds themselves, build farmers' technical expertise, or develop new products for niche markets, like organically grown food.

PLANT MUTATION

It may be asked now how a plant comes to have two aspects of a character or how a character pair comes to be. It is known that the wild-type gene as it occurred originally may alter its nature. The alteration produces a new expression of the character.

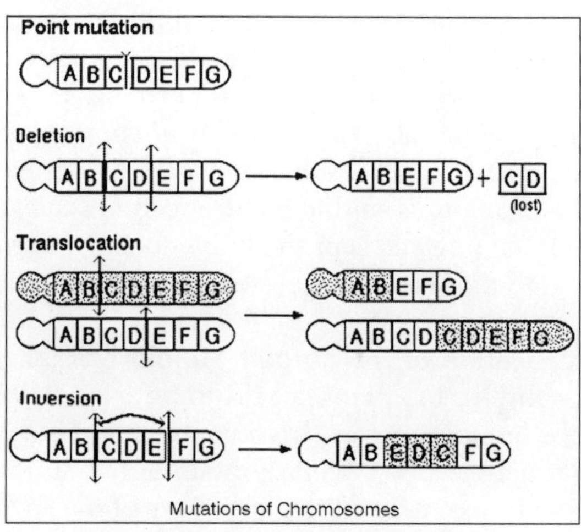

Fig. Mutation

The process, or act of alteration, is called mutation. The new type of plant so produced is called a mutant or a mutation. A gene in the majority of cases mutates only once, and thus gives a character pair, the wild type and the mutant. The same gene, however, may mutate again and give three forms of the gene—the original wild type and two mutants. This provides three expressions of the character.

Only two of these three can be present in any diploid somatic nucleus, and only one in any gamete. A case may be taken from the haricot bean, plants of which may have green foliage and green pods (Gr) or green foliage and yellow pods (G), or yellow foliage and yellow pods (g). When Gr is crossed with G, the Fi shows green pods and foliage. The Gr condition is dominant to the G condition. When the G type is crossed with the g, the green foliage and yellow pods condition is dominant.

Thus we see a series of decreasing dominance—Gr dominant to G, and either of these dominant to g. Such a graded set is called a multiple-allelomorphic series. All sorts of quantitative and physiological conditions are known where the genes for the characters belong to such a series. In some series there may be more than three aspects, for cases are known where the gene has mutated five or six times, and six or seven character expressions exist.

ASPECT OF CYTOLOGY IN HETEROPLOID

Another aspect of cytology affecting Mendelian ratios must now be referred to. It sometimes happens that at mitotic or meiotic division the daughter nuclei do not separate cleanly in the usual manner. In these cases the chromosomes, either as a whole set or as part of a set or even individual members, lag behind. The laggards which should have gone to one daughter nucleus are included in the other. When this occurs, nuclei are formed with a number of chromosomes different from that normally found in the species.

A nucleus with such an aberrant number of chromosomes or a plant in whose nuclei they occur is said to be heteroploid. If a whole chromosome set is replicated, the heteroploid is said to be balanced or *euploid.* If only a part of the set is replicated, the nucleus is of the unbalanced or *aneuploid* class. Most attention is directed to those cases where whole sets are concerned, that is, to the euploid form.

Where one replication of the normal diploid has taken place, and the homologues are in threes, the plant is said to be *triploid.* Where the diploid has doubled and the homologues are in fours, the plant is said to be *tetraploid.* Plants are known where the body cells contain only half the normal diploid number; that is, the number normal to the gamete of the species. These plants are known as *haploids.*

There is thus an ascending series in the terminology: haploid, diploid, triploid, tetraploid, pentaploid, hexaploid, and upwards. These have been formed in nature as the result of some unknown stimulus, or artificially in cultivation by the use of such drugs as colchicine.

HETEROPLOID'S EFFECT

Giantism

Replication of chromosome sets often results in the production of a bigger (giant) form of plant. This in itself may be very useful as, when it occurs in a plant of a forage crop.

Sterility and Fertility

Heteroploidy may be very useful in regard to sterile hybrids. These hybrids often result from crosses between plants which are not very closely related, but close enough to form a new hybrid plant. The chromosomes derived from the two parents are sufficiently homologous to form a mixed nucleus in Fi, but not so as to behave normally at meiosis. The hybrid fails to form viable gametes.

Such hybrids or "mules"are sterile, and therefore, unless they can be propagated successfully and economically, are of little or no use in practice. Replication of the mixed sets of chromosomes, if it occurs in such a sterile hybrid, may render it fertile.One of the clearest examples of this is furnished by a non-agricultural plant.

Primula verticillatay with nine pairs of chromosomes in the somatic nuclei, can be crossed with Primula floribunda, which also has nine pairs of chromosomes.The gametes of each, therefore, contain nine chromosomes, and these form the hybrid. None of the nine floribunda chromosomes are homologous with the nine verticillata members, and therefore successful meiosis in the hybrid is barred.

Viable gametes are not possible, and the plant is sterile. The sterile hybrid formed a euploid and immediately became fertile. Schematically, the mating, the chromosomes being symbolized by the initial letter of the species from which they are derived.By replication, each V chromosome obtains a homologue and similarly each F chromosome. Normal meiosis is possible and gamete production follows. Many species not closely related can form hybrids which are sterile.

It is probable that many of these will be rendered fertile when duplication occurs. Wide crosses as between, a wild grass like couch and wheat, or between turnip and cabbage, are important because the new gene complex brought into being results in more than a mere addition of the characters of one parent to those of the other. Genes even in a normal diploid soma interact on each other to some extent, and when two complete gene complexes interact the effect is greatly increased, so that a new plant different from both parents is brought into being.

This is paralleled in chemistry when, the union of sodium, a white silvery metal, with chlorine, a green gas, produces a new substance, sodium chloride with characters quite different from the parent substances. The formation of heteroploids has occurred in nature. Our modern bread wheat as it was evolved is a complex hexaploid of diverse parentage with three sets of seven, that is, twenty-one, chromosomes in the gamete, and twenty-one pairs (forty-two individuals) in the somatic nucleus. One interesting result of heteroploidy occurs when the component sets of chromosomes each contribute to the new plant similar genes.

A case in point arises in connection with the production of the red colour in the grain of bread wheat. Here, if an appropriate variety with red grain is crossed with a variety having white grain, a 3: 1 ratio red. If a different red-grained variety had been used as a parent, a 15: 1 ratio red: white in Fs would have been obtained. If yet another variety had been employed, the ratio of red: white in F2 would have been 64: 1. Results such as these indicate that three pairs of genes are involved in the production of grain colour.

Furthermore, each pair must be located on different chromosome pairs, because the genes segregate and recombine freely. In short, there must be six chromosomes (three pairs) in each hexaploid somatic nucleus, all of them carrying a grain-colour gene.

Cytological evidence shows the bread wheat to be hexaploid. Multiple genes (polymeric series) of this type become important when they affect quantitative characters, and each dose of, the dominant produces an increase in the character. Sugar content of sugar beet depends on a set of replicated genes, and every additional "dose"of gene produces an increase of sugar content.

SELF-FERTILIZATION AND CROSS-FERTILIZATION IN STERILITY

Self-fertilization and cross-fertilization have been referred to. These are often barred by sterility. Almost any cabbage plant will illustrate the point. If its own pollen only is allowed access to its stigmas a cabbage will set few or no seeds. If pollen from other cabbage plants gets access to the stigma, fertility is complete.

This proves that its ovules are fully functional. Its pollen is fully effective in producing seed on most others of its fellows. The gametes of the plant accordingly are fully viable but self-incompatible.

It is now known that the tube of a pollen grain carrying any Sallelomorph can hardly pass through a style carrying this same allelomorph in the pair resident in the cells. A similar mechanism affects many varieties of fruit tree. If an orchard has been planted only with such a self-incompatible variety, every one of the trees produced by vegetative propagation will have exactly the same gene complex. Hence, every stigma will contain incompatible genes in common with the pollen supply and fertility will be barred. Such an orchard must be interplanted with a few trees of another variety which is known to be capable of supplying suipollen.

Many plants are known to have a cytological constitution which does not permit of normal meiosis. The plants are usually of hybrid origin. Many fruit trees are of this nature and do not form viable gametes.

VIGOUR LOSS

Where sterility is no bar, and continued self-fertilization is practised, another disability may show itself. The vigour of the progeny derived through continued self-pollination falls progressively generation by generation. In maize, after a few generations of self-fertilization the inbred lines become weak, and in extreme cases may fail completely and die out.

Hybrid Vigour

Curiously, when two inbred but unrelated lines are crossed the F1 is often extremely vigourous. Out-breeding brings back vigour. This vigour in a hybrid has been called *heterosis* or *hybrid-vigour*. It shows only in the F1 and does not appear in F2 when the F1 is self-fertilized.

Fig. Heterosis

When appropriate inbred lines are used as parents, the Fi is often very much more vigourous and yields much more than any commercial lot. Heterozic or Heterozygous maize seed is now sold for crop production. In fact, for this crop, practically all progressive growers sow only hybrid vigourous seed. All of the Fi crop goes for consumption as food, etc.; none is retained for sowing. Each year the Fi is made anew from inbred parental lines which are known to give the maximum effect when crossed.

The commercial production of heterozic maize seed is simple. The two parental lines are interplantcd in rows. At flowering time all the male inflorescences are removed from the plants designed as the seed parent. It is therefore pollinated only by its neighbour, and all the "seed"produced by it is of hybrid origin. In any crop produced by vegetative propagation the vigourous hybrid need be produced only once. All clones arc genetically identical and just parts of the Fi itself. They retain all its characteristics in full effect. This is well seen in the poplar tree used for wood pulp. Two parents arc crossed. If the progeny is highly vigourous, suckering is induced. The suckers are planted out. The trees from such suckers grow very fast and harvest is brought nearer by a number of years.

SELECTION OF PLANT

Cytological aberration and genetical hybridity provide plant populations variable in character. Variation in a crop, no matter how induced, provides the plant selector with his material. Two inbred strains of Maize on eitiier side of the vigourous hybrid got from their mating. A new combination of genes or gene complexes can be the starting point of a new variety waiting to be selected and multiplied. There are three methods of selection.

Pure Lines

The first consists of selecting single plants, self-fertilizing them, and maintaining the progeny of each as a separate "line." In effect this subjects each selection to the progeny test.

The best line which breeds true can be retained and the others scrapped. This method is known as single-plant selection. If the original parent was homozygous for the characters observed, a non-varying or *pure line* will have been formed.

A pure line consists of the descendants of one homozygous individual, self-fertilized and never cross-fertilized.

This amounts to *genotypic* selection. A line pure for *all* genes is probably not possible, but a line pure in all the genes that matter is obtainable in practice. It is to be noted that, so far as concerns the genes for which the line *is* pure, further selection can have no effect. A pure line, then, by definition is non-varying.

Apart from environmental fluctuation and the possibility of mutation, no alteration in appearance or behaviour is possible.The second method of selection is to choose all the plants in a crop which appear to conform to a predetermined standard. Such is mass or *phenotypic* selection. This is often useful with plants which are individually self sterile, or where the character is quantitative and depends on a comparatively large series of polymeric genes.

Mass selection gives an immediate effect with a large bulk in the first generation. As generation follows generation, if re-selection is not practised on each, the effect wears off, and the strain "runs back"to an average. Mass selection is often possible on large-scale lines. In many countries growing crops are inspected officially, and where a crop conforms to a definite standard the grower receives a certificate for it.

When such a certification scheme is widely patronized it amounts to a large-scale mass selection.

The third method of selection is not to select out the good plants, but to pick out from a crop the wrong ones or "rogues."

These are the plants which do not conform to standard. Roguing has a potent influence when the rogue is due to a dominant gene, and when the desired character is due to a recessive.

Roguing out a recessive is not really effective in practice, for it leaves in the crop the heterozygotes. In this latter case it usually pays better to develop a pure line following a progeny test.

CLASSICAL PLANT BREEDING FOR PEST RESISTANCE

Wild relatives of crop plants such as beans, wheat, and maize are not uniformly resistant to insect and disease pests. This can be demonstrated in simple fashion—when selections of these wild populations are set out in plant-rows, some of them are highly susceptible, others are resistant, and some are intermediate in resistance to the common pests of the region.

The first plant breeders, those women and men who domesticated crops such as beans, maize, and wheat, could save only those genotypes that had some level of resistance, *i.e.*, those individual plants that did not succumb to pest depredation. In effect, therefore, they selected for pest resistance and thus changed the population structure of their crop species in favour of resistance genes. This change made it possible to grow the crops in monoculture, which was convenient for food production and harvest. It was also convenient for multiplication of disease and insect pests that might not be affected by the limited sample of resistance genes.

Plant breeding thus set the stage for sequential cycles of pest resistance and pest susceptibility of crop plants. We have no direct record of the consequences of this ancient ecological meddling, but myth and historical accounts tell of disastrous disease epidemics and insect outbreaks, so one can assume that from time to time large plantings of crops that were uniformly susceptible to a new kind of insect or disease fostered increases of that pest to epidemic proportions. Resistance genes were essential for crop domestication and monoculture but they did not guarantee perfect safety. We have no record at all and little or no speculation about how the newly domesticated crops might have affected their wild relatives, which no doubt were growing in close proximity to the domesticates.

RESISTANCE BREEDING AFTER MENDEL

Genetics-based plant breeding, launched in the early years of the 20th century, produced new crop varieties with improved resistance to major disease and insect pests. Usually such resistance was developed as a second phase—a rescue operation—after new varieties, selected primarily for high yield, were discovered to be susceptible to a particular insect or disease. Breeders found early on that they could identify single genes (usually dominant) that conferred essentially complete resistance to the pest in question. Varieties containing such excellent resistance were developed and released for large-scale farmer use. But breeders then discovered, all too often, that the "perfect" resistance lost its effectiveness after a few seasons. They soon learned, with the aid of entomologists and plant pathologists, that insect and disease pests are highly diverse genetically, and that almost without fail a rare pest genotype will turn up (or perhaps be created *de novo* by natural mutation) that is not affected by the newly-deployed resistance gene. The new pest genotype multiplies and the crop variety's resistance "breaks down."

As years went by, breeders found that some kinds of resistance did not fail, and that such resistance often was less than complete; the plants suffered some damage but gave satisfactory performance overall. This longer lasting resistance was dubbed "durable" resistance. Further, the breeders discovered that durable resistance usually (but not always) was governed by several genes rather than by one major gene. The multifactorial kind of resistance has been called "horizontal resistance." The major-gene resistance has been called "vertical resistance."

The good news, then, was that breeders could identify and breed for durable resistance. The bad news was that the breeding was more difficult because several genes had to be transferred at one time, thus requiring larger populations for selection, as well as multiplying the usual problems with "linkage drag" (undesirable genes that are tightly linked to the desired ones). To this day, breeders use both kinds of resistance in varying proportions, according to the crop and where it is grown. At first, breeders found and used resistance genes from the adapted, local landrace populations that also were the initial gene pool as a source of resistance genes for their new varieties. As years went by, these gene pools began to dry up and breeders looked further afield, turning to exotic (unadapted) landraces, and even to wild relatives of their crop. Sometimes they made extraordinary efforts to hybridize the domestic crop with a very distant wild relative—making a cross that could not succeed under natural conditions. Embryo rescue and even x-ray treatments were used to make "unnatural" crosses and derive breeding progeny from them. The breeders fooled around with Mother Nature; they moved genes farther than natural processes would allow.

But the breeders as a whole preferred to not breed from exotic varieties or distant and often wild relatives. They used exotic material only when there was no other choice. This preference was due not only to the difficulty of wide hybridization, but also to the fact that exotic germplasm exacerbates the problem of undesirable linkages. Few or none of the foreign genes—except the desired resistance genes—were suitable for the needs of high yielding, locally adapted varieties. But often the breeders had no choice; either they got the needed resistance genes from a distant relative, or they got nothing at all.

At about this time, breeders realized that it would be important to conserve remnant seed of landraces from all around the world, but especially from the centres of diversity of their crop. As farming worldwide grew more commercial, farmers turned more and more to professionally bred varieties that were better suited to commercial production, and in so doing they abandoned their landraces. If remnant seed of those landraces was not collected and saved in special storage facilities, the genetic base for crop breeding in the future would be drastically narrowed. Seed "banks" were needed. Through the efforts of a few far-sighted plant breeders, seed banks were established in several countries and in international research centres.

So at the end of the 20th century, plant breeding for pest resistance had laid out the genetic framework of vertical and horizontal resistance, and identified important sources of new resistance genes, *i.e.*, plant germplasm from anywhere in the world. Sources were limited, however, to the crop species itself or its relatives, either wild or cultivated. All of the introduced genes therefore came from plants. Plant breeders selected not only for tolerance or resistance to disease and insect pests, they also selected for tolerance to abiotic stresses such as heat and drought, cool temperatures, or nutrient imbalance. Much of this selection was involuntary; in selecting varieties with top performance over many seasons and many locations the breeders necessarily selected varieties with tolerance to the prevailing abiotic stresses of the diverse seasons and localities. In selecting for tolerance to environmental stresses, breeders necessarily changed the genetic makeup of the crop species, altering it still further from that of the original wild species, which had been restricted to certain environmental niches. Witness teosinte (the probable parent of maize), restricted to certain habitats in Mexico as compared to maize that now is grown in nearly every country of the world except Iceland. Global distribution of crop plants often means that they are grown with no proximity to wild relatives that might intercross with them. Teosinte is not found in Germany or China, nor for that matter in the US Corn Belt. In other cases, however, wild species with hybridization potential coexist with their cultivated crop relatives, often as weeds. Canola, sunflower, and grain sorghum are examples of crops with hybridization potential with either a related species (canola with wild mustards) or with a weedy form of the same species (sorghum with shattercane, cultivated sunflower with wild sunflower).

REASONS FOR PRIVATISATION OF PLANT BREEDING

Plant breeding can be conceptualised as an investment that develops improved varieties with the potential to generate future benefits in the form of improved crop productivity, reduced costs of production, and/or higher returns. Potential value from improved cultivars will be realised only if and when farmers adopt these cultivars in their cropping systems, AND when consumers willingly purchase the food or other crop products in a competitive market. Growers will only adopt these new varieties if they provide real financial benefits that exceed the costs of adoption, including any additional costs of acquiring the improved variety.

Similarly, consumers will only knowingly purchase food from these new varieties if by so doing they derive a net benefit in the form of enhanced attributes and/or lower prices relative to available alternatives. In common with other forms of investment, the rate of return will depend on the discounted value of the flow of future benefits net of present value of all costs neccessary to generate such benefits. Arguably the most important reason for the growing trend to privatisation of plant breeding has been significant changes in the ability of plant breeders to appropriate at least some of the benefits from improved grain varieties that otherwise would be captured by growers.

Specifically, the incentive for firms to invest in plant breeding depends on the ability to exclude grain growers, and often competing plant breeders as well, from commercial exploitation of a breeder's varieties unless they pay to do so.

For some crops such as corn, the development of hybrid technology provided genetic copy protection that enabled plant breeders to capture much of the value from heterosis as well as other superior traits. For other crops, it has been the expanding scope of intellectual property rights that has enabled the capture of some of the value created by plant breeding. So while it has been the application of modern science to plant breeding that has generated much of the potential for value creation in the grain supply chain, it has been extensions to the legal framework for intellectual property rights that have made possible private capture of enough of the value created by plant breeding to provide the private sector with an incentive to invest more in plant breeding.

The most significant of these intellectual property rights are patents and Plant Breeder's Rights. In recent decades, both court decisions and legislative changes have expanded the scope and impact of these two types of intellectual property right appreciably.

Complementing these developments in the institutional framework have been scientific discoveries that led to greater potential for value creation by improvements in plant breeding methods. Apart from hybrid technology, other recent advances include new technologies that improve the efficiency of all plant breeding, including both conventional and transgenic plant breeding.

Such techniques include double haploidy, plant regeneration systems, molecular based hybrid technologies, and marker assisted selection. Use of these techniques in conventional plant breeding is already reducing the time lags from initial crosses to release of new varieties. Potentially beneficial outcomes from the application of these technologies to plant breeding include one or more of the following:

- cheaper development of improved crop varieties.
- faster/earlier development of improved crop varieties.
- development of superior improved crop varieties, that are more productive, produce better quality grain, or both.

In addition, there has been the more controversial development of transgenic technologies used to produce GMO's. Potential beneficial outcomes from transgenic technologies include:

- development of improved crop cultivars with novel agronomic/ input traits that enable lower average costs of production.
- development of improved crop cultivars with novel quality-enhanced traits for which consumers are willing to pay a price premium.

On the other side of the coin, publicly funded rural research has been under pressure for at least the last two decades. In part, this been due to a growing perception that grain growers have been the primary beneficiaries of the traditional plant breeding programmes.

Historically these programmes have been funded mainly from consolidated revenue. To some extent, this concern has been addressed by the relatively recent evolution of the GRDC and similar bodies that rely heavily on collective industry funding to support much of their investments, but the fact remains that a significant part of plant breeding programmes is still publicly funded.

Governments also now demand greater accountability at the same time that they reduce funding for agricultural research and extension. As a result, many "public" institutions are under pressure to become at least partially self-funding, and are starting to charge for selected goods and services. Public research institutions also seek to patent and/or commercialise discoveries made in the course of government funded research, or pursue opportunities to license technologies to the private sector. Public plant breeding programmes have not been immune to government pressure to generate revenue from their activities. Like private business, their capacity to capture a high proportion of the net benefits of new varieties depends on:

- a legal basis to establish ownership of the intellectual property embodied in the variety,
- the capacity to exclude potential users who are not willing to pay the nominated price,
- the costs of monitoring and enforcing compliance,
- the capacity for price discrimination.

Pricing practice by public institutions is still evolving. If they start charging significant fees at levels approaching full cost recovery, and exclude farmers unwilling to pay these fees from access to new varieties, then they cease to be public plant breeding organisations within the meaning of the term in this chapter.

Finally, agronomic practice by grain growers has become increasingly sophisticated and much more tactical. In particular, many growers now make decisions about which varieties to grow each season on the basis of the latest possible information about the climatic outlook and other seasonal indicators, such as soil moisture levels as well as weed and disease threats. Consequently they are less likely to use seed saved from the previous harvest, and more likely to purchase new seed of the desired variety from a seed merchant. This change in farming practice will increase the size of the seed market, and improve the economics of private plant breeding.

VALUE CREATION AND CAPTURE ISSUES

It is inevitable that the growing privatisation and commercialisation of plant breeding in the Australian grains industry will lead to increased competition between plant breeders. While this trend is seen as threatening by some people, there are compelling grounds for viewing this change as an opportunity to create more value for the industry and the nation provided that certain conditions are met. First though the way in which plant breeding creates value is discussed.

In a market economy, the ultimate determinant of the aggregate value of all activities in the food supply chain is the amount that consumers are willing to pay to consume end-products. Value creation occurs when there is an improvement at any step in the supply chain that increases aggregate value in consumption.

Market forces, mediated by the institutional and policy framework, including intellectual property rights, will determine the extent to which value is captured at various stages along the supply chain. In the grains industry, the potential exists for new varieties to create value by lowering the cost of producing and delivering grain and grain products to consumers; and/or by enabling the production of superior grain products for which consumers are willing to pay higher prices. This potential for value creation will be realised when new varieties from the breeding programme are released and adopted by grain growers, and the resulting products are purchased and consumed by end-users.

The application of modern science to plant breeding has dramatically increased the potential to create extra value in the grain supply chain. New plant breeding methods such as di-haploidy, embryo rescue, and rapid breeding cycles have sped up the development of new varieties and reduced breeding costs. Furthermore, molecular marker technology enables breeders to be much more selective and effective at identifying desirable traits in germplasm collections and incorporating these traits into elite lines. Transformation technologies have significantly expanded the range of traits that plant breeders can access.

Greater value will be created if plant breeders can produce more and better varieties faster and cheaper. To fully realise this potential for value creation, more rather than less investment in plant breeding is needed, and plant breeders need to take a more selective and strategic approach to plant breeding decisions.

For reasons already discussed, less rather than more public funds are likely to be available for investment in plant breeding, so most new investment will need to be privately funded. Funds will be forthcoming only if prospective rates of return from private plant breeding are sufficiently attractive, and this will depend on the extent to which market forces and intellectual property rights allow and enable breeders to capture the value that they create. Innovation involving changes to established practice in plant breeding is essential to take full advantage of the availability of the new breeding technologies.

Sometimes organisations are able to achieve such a transformation in the absence of competitive pressures, but often they cannot. For both public sector organisations and private firms, competition commonly provides the necessary impetus to drive change and risk taking.

Notwithstanding the potential benefits from privatisation, some people perceive threats to the interests of grain growers and/or to the national interest. Some threats are more imagined than real. Other threats are real enough for some individuals or groups, but are outweighed by broader community benefits. There also are some legitimate grounds for concern about institutional arrangements that could limit the efficiency gains from a more competitive plant breeding system.

One possible outcome of greater privatisation of plant breeding is that a small number of large multi-national firms crowd out other competition, and that they might then exploit their market power to capture almost all of the value created by plant breeding.

Pricing practices by these large life science companies also might threaten future competitiveness of other sectors of the farming industry, as well as limiting widespread value creation. The commercialisation of Bt cotton by Monsanto in the USA and in Australia provides one such example. Bt cotton was commercially introduced in the USA in early 1996, and later the same year in Australia in time for the 1996/97 growing season. Monsanto Corporation is the patent holder for the Bt technology, and sought to appropriate the benefits generated by its intellectual property through a technology fee. In the USA, the initial level of the technology fee charged to cotton growers was US $32/ acre.

In Australia, Monsanto set the technology fee at A$245/ha, which at then current exchange rates translated to about US $70/acre, or approximately double the level set in the USA. Even though Monsanto subsequently introduced some concessions in response to pressure from the domestic industry, Australian cotton growers still ended up paying a considerably higher technology fee than their US counterparts. The considerable monopoly power that the large life science companies possess over proprietary key enabling technologies for plant breeding also is a possible threat to continued technology development in a country such as Australia. Arguably, legal disputes over intellectual property rights that block widespread utilisation and/or further development of the technology pose a major threat to long run value creation from transgenic technologies. Concerns have been expressed about possible patent gridlock, excessive secrecy and duplication of inventive effort, excessive transaction costs to license patented technology, prisoner dilemma type impasses, and/or the "tragedy of the anti-commons".

Such supply side problems might tie up the technology in the courts and block commercial implementation for years, if not decades. It also may threaten the *freedom to operate* for public and local industry research and plant breeding programmes.

Given examples such as this, it is not surprising that Australian grain growers have already demonstrated a willingness to fund local plant breeding firms to forestall such a threat, and to ensure ongoing access on reasonable terms to locally bred varieties that maintain Australia's competitive advantage in international grain markets.

Finally, there are real concerns relating to the provision and utilisation of certain key "knowledge rich" inputs to the plant breeding process.

Examples of these "knowledge rich" inputs include germplasm collections of land-race and elite breeding lines, and results of pre-breeding research such as agronomy; biometry; entomology; genetic mapping and molecular marker development; germplasm collection and conservation; information and database systems; molecular biology research; plant pathology; plant physiology; plant quarantine; product chemistry; and quantitative genetics.

Most such inputs are key enabling technologies for plant breeding that provide the foundation for ongoing long-term variety improvement and consequent productivity gains.

Effectively, they also are non rival in use by competing plant breeders because they are largely knowledge based.

Where knowledge is embedded in a tangible technology, the use of which requires consumables that are rival in use, the knowledge component of such technology is still non-rival in use. In effect, the knowledge component enables a capacity to practice the technology that is unlimited, and hence non-rival in use. Hence they share at least one key attribute of a public good, in so far as use by any one plant breeder does not diminish the value of the input to other plant breeders.

This capacity to practice key enabling technologies will be referred to as essential plant breeding infrastructure (EPBI). Traditionally, such inputs were non-proprietary, provision was publicly funded, and access by public plant breeding programmes was free and open.

In return, no attempt was made to recover the costs of the breeding programme (as distinct from costs of seed multiplication) by charging growers for the intellectual property embodied in newly released varieties. In a world of privatised plant breeding, these admirable arrangements are unlikely to survive. Current funding sources for the provision and further development of key enabling technologies already are under threat. As a result, public agencies may abandon such activity to the private sector, and/or may seek to recover some or all of the cost by charging plant breeders for access to these technologies. This is likely to hasten an emerging trend towards greater application of intellectual property rights to breeding technologies as well as to germplasm, and to the commercialisation and possible privatisation of their production.

If key enabling technologies are proprietary or otherwise price excludable, then they belong to a class of goods referred to as excludable public goods. This may provide the necessary incentive for private sector investment to compensate for declining public and/or collective industry funding.

If it does not, eventually innovation and consequent returns to private investment in plant breeding are likely to stall as a result. However, the stimulus to inventive activity provided by the patent system in other areas demonstrates that private provision of excludable public goods is feasible, and may have advantages vis-Å-vis public provision of pure public goods.

Nonetheless, there also are threats to the efficient provision and utilisation of EPBI that need to be addressed if the potential benefits from some form of privatised provision of key enabling technologies are to be fully realised. For instance, if private firms respond individually rather than collectively, there will be wasteful duplication of effort in producing such inputs as commercial plant breeders strive for competitive advantage in the market place. Moreover, there is the risk that the incentives for an individual private investment will be inadequate for it to substitute fully for the likely withdrawal of public funding.

Either of these two scenarios is inefficient. The obvious solution is for cooperative behaviour to provide such inputs. This may involve joint funding by private plant breeders, such as international consortia formed to develop molecular marker technology.

However, given the history of funding of plant breeding, the more likely scenario for most EPBI is a continuation of collective funding by industry through a body such as the GRDC. Note that while "membership" of GRDC is compulsory for all producers of mandate grains, the cost of membership is linked to the level of grain production. Nevertheless, there is increasing pressure for GRDC to commercialise those enabling technologies in which it invests, and to recoup at least part of the cost of its investment by way of user fees.

Irrespective of whether funding for future provision of essential plant breeding infrastructure comes for collective funding or private investment, it is highly likely that there will be a lack of competition in the provision of essential plant breeding infrastructure. While adequate and efficient provision of EPBI is one cause for concern, efficient utilisation is another. These inputs are quasi public goods in the sense that they are non-rival in use. As a result, efficient utilisation involves both the potential for under-utilisation of essential plant breeding infrastructure due to monopoly pricing, and the much maligned concept of the "level playing field" in downstream industries.

Economists' interest in this concept stems from the observation that if the institutional, policy, or legal framework confers advantages on some firms relative to others, competition may not generate desirable outcomes. If the favoured firms are not the most efficient, the outcome may be that inefficient producers out competing their more efficient counterparts. Conversely, if all firms compete on a "level playing field", then the law of the jungle should ensure that only the most efficient survive.

There are obvious parallels here to National Competition Policy (NCP) principles governing access to essential infrastructure. The aim of National Competition Policy (NCP) is to facilitate effective competition where competition between suppliers of goods and services result in lower prices, a wider range of products, and/or better service for consumers, but also to accommodate situations where competition does not have that effect, or where it conflicts with social objectives.

In industries such as telecommunications, air and rail transport, and electricity transmission, NCP recognises that competition may not be feasible or desirable in the provision of some essential infrastructure, and that the shared use of such 'bottleneck' or 'essential' infrastructure facilities may be necessary to facilitate efficient competition in downstream markets that use such infrastructure.

Access regulation that aims to promote competition in markets that use the services of 'essential' infrastructure while preserving incentives to develop and maintain those facilities have been developed to address concerns about denial of access and/or monopoly pricing of access.

A case can be made that, as plant breeding becomes increasingly privatised, equivalent access regimes will need to be developed for essential plant breeding infrastructure (EPBI).

Unless such an access regime is established, some of the potential benefits from scientific discoveries underpinning modern plant breeding may not be fully realised. In common with NCP access regimes, the aim should be to promote full and efficient competition between plant breeders, while preserving adequate incentives for investment in the ongoing development, maintenance, and provision of essential plant breeding infrastructure. GRDC, as the key provider of EPBI for crop breeding, is cognisant of this problem, and is working to develop policies for access by all plant breeders to essential plant breeding infrastructure in which it invests.

A key issue that will need to be addressed in these policies is the grounds (if any) for denial of access or for discriminatory pricing. For example, one possibility would be to deny access to, or charge higher prices for EPBI to large multi-national "life science" firms. A possible ground for doing so would be that these multi-national firms have access to other sources of EPBI from which Australian plant breeding firms are excluded. Another possible ground would be that GRDC has invested in selected plant breeding firms. Fears have been expressed that they may decide to "protect" such investments by limiting other plant breeders' access to GRDC funded EPBI. *Prima facie,* denying access or discriminatory pricing for this reason would seem to be an example of exploiting market power in order to benefit owned or related entities in upstream or downstream markets. Specifically, it would inhibit rigorous competition in the downstream plant breeding market. Nevertheless, there may be grounds for treating overseas owned plant breeding firms differently to domestically owned firms because of the potential impact on Australia's trading position.

2

Plant Tissue Culture

PLANT TRANSFORMATION

Practically any plant transformation experiment relies at some point on tissue culture. There are some exceptions to this generalization, but the ability to regenerate plants from isolated cells or tissues *in vitro* underpins most plant transformation systems.

PLASTICITY AND TOTIPOTENCY

Two concepts, plasticity and totipotency, are central to understanding plant cell culture and regeneration. Plants, due to their long life span, have developed a greater ability to endure extreme conditions and predation than have animals. Many of the processes involved in plant growth and development adapt to environmental conditions. This plasticity allows plants to alter their metabolism, growth and development to best suit their environment. Particularly important aspects of this adaptation, as far as plant tissue culture and regeneration are concerned, are the abilities to initiate cell division from almost any tissue of the plant and to regenerate lost organs or undergo different developmental pathways in response to particular stimuli. When plant cells and tissues are cultured *in vitro* they generally exhibit a very high degree of plasticity, which allows one type of tissue or organ to be initiated from another type. In this way, whole plants can be subsequently regenerated. This regeneration of whole organisms depends upon the concept that all plant cells can, given the correct stimuli, express the total genetic potential of the parent plant. This maintenance of genetic potential is called 'totipotency'. Plant cell culture and regeneration do, in fact, provide the most compelling evidence for totipotency.

The Culture Environment

When cultured *in vitro*, all the needs, both chemical and physical, of the plant cells have to met by the culture vessel, the growth medium and the external environment (light, temperature, etc.). The growth medium has to supply all the essential mineral ions required for growth and development. In many cases (as the biosynthetic capability of cells cultured *in vitro* may not replicate that of the parent plant), it must also supply additional organic supplements such as amino acids and vitamins.

Many plant cell cultures, as they are not photosynthetic, also require the addition of a fixed carbon source in the form of a sugar (most often sucrose). One other vital component that must also be supplied is water, the principal biological solvent. Physical factors, such as temperature, pH, the gaseous environment, light (quality and duration) and osmotic pressure, also have to be maintained within acceptable limits.

PLANT REGENERATION

Having looked at the main types of plant culture that can be established *in vitro*, we can now look at how whole plants can be regenerated from these cultures.

In broad terms, two methods of plant regeneration are widely used in plant transformation studies, *i.e.* somatic embryogenesis and organogenesis.

SOMATIC EMBRYOGENESIS

In somatic (asexual) embryogenesis, embryo-like structures, which can develop into whole plants in a way analogous to zygotic embryos, are formed from somatic tissues. These somatic embryos can be produced either directly or indirectly. In direct somatic embryogenesis, the embryo is formed directly from a cell or small group of cells without the production of an intervening callus. Though common from some tissues (usually reproductive tissues such as the nucellus, styles or pollen), direct somatic embryogenesis is generally rare in comparison with indirect somatic embryogenesis.

In indirect somatic embryogenesis, callus is first produced from the explant. Embryos can then be produced from the callus tissue or from a cell suspension produced from that callus.

Somatic embryogenesis usually proceeds in two distinct stages. In the initial stage (embryo initiation), a high concentration of 2,4-D is used. In the second stage (embryo production) embryos are produced in a medium with no or very low levels of 2,4-D.

In many systems it has been found that somatic embryogenesis is improved by supplying a source of reduced nitrogen, such as specific amino acids or casein hydrolysate.

Organogenesis

Somatic embryogenesis relies on plant regeneration through a process analogous to zygotic embryo germination.

Organogenesis relies on the production of organs, either directly from an explant or from a callus culture. There are three methods of plant regeneration via organogenesis. The first two methods depend on adventitious organs arising either from a callus culture or directly from an explant. Alternatively, axillary bud formation and growth can also be used to regenerate whole plants from some types of tissue culture. Organogenesis relies on the inherent plasticity of plant tissues, and is regulated by altering the components of the medium. In particular, it is the auxin to cytokinin ratio of the medium that determines which developmental pathway the regenerating tissue will take. It is usual to induce shoot formation by increasing the cytokinin to auxin ratio of the culture medium. These shoots can then be rooted relatively simply.

Tissue culture and plant regeneration are an integral part of most plant transformation strategies, and can often prove to be the most challenging aspect of a plant transformation protocol. Key to success in integrating plant tissue culture into plant transformation strategies is the realisation that a quick (to avoid too many deleterious effects from somaclonal variation) and efficient regeneration system must be developed. However, this system must also allow high transformation efficiencies from whichever transformation technique is adopted.

Not all regeneration protocols are compatible with all transformation techniques. Some crops may be amenable to a variety of regeneration and transformation strategies, others may currently only be amenable to one particular protocol. Advances are being made all the time, so it is impossible to say that a particular crop will never be regenerated by a particular protocol. However, some protocols, at least at the moment, are clearly more efficient than others. Regeneration from immature embryo-derived somatic embryos is, for example, the favoured method for regenerating monocot species.

TECHNIQUES IN TISSUE CULTURE

Fungi and bacteria grow on the plant surface and contaminate the culture media when they are not adequately eliminated. The process of explant introduction into in vitro conditions depends mainly on the disinfecting phase. Problems in handling produce contamination of the in vitro plant and so, a similar disinfection process to that of the in vitro introduction technique is recommended.

However, it is possible to use antibiotics included in the medium, but just temporarily, while the plantlets are growing. Among the bacterial contaminants are Bacillus sp., Erwin/a sp., Pseudomonas sp., etc. In the case of systemic infections, the use of disinfectants is not effective, because the pathogens are located in the vascular system. In this case, it is recommended to use specific antibiotics and meristems or bud cuttings.

On the other hand, the process of in vitro introduction involves the use of explants in different physiological stages, as dormant buds, which require growth regulators for example, Gibberellic acid) to stimulate and accelerate bud growth. During the process of in vitro introduction there is a close connection between the greenhouse and the laboratory; that is why the necessary precautions should be taken to avoid the entrance of contaminants into the laboratory.

- In the greenhouse
  - Take cuttings from the mother plant, or buds from greenhouse tuber sprouts.
- In the laboratory
  - Immerse the cuttings for 10 minutes in a beaker conveniently labeled with the accession name, and containing a solution of 0.5 per cent acaricide with three drops/I of Tween 20.
  - Throw away the acaricide solution and rinse the nodes with tap water
  - Prepare the laminar flow chamber.
  - Immerse the cuttings in 700/o alcohol for 30 seconds and take the beakers immediately to the laminar flow chamber
- In the laminar flow chamber
  - Eliminate the alcohol and replace it with a 2.50/o solution of calcium hypochlorite. Keep the cuttings immersed during 15 minutes.
  - Throw away the hypochlorite, and rinse the cuttings three times with sterile water
  - Keep the nodes immersed in sterile water until bud extraction.
  - Dissect the buds on a sterile plate and place them in the culture medium.

MICROPROPAGATION

In vitro plantlets, which are free of pathogens, are used as initial material for potato and sweetpotato seed programmes. The methods used in these micropropagation programmes mainly depend on their production volume and the available infrastructure.

In the case of potato micropropagation, the basic methods have been already described. They have been verified in many institutions and they are based on the rapid growth of individual node cuttings, or stems with multiple node cuttings. Afterwards, the basic micropropagation methods are described.

NODE MICROPROPAGATION

This method is based on the principle that the node of an in vitro plantlet placed in an appropriate culture medium will induce the development of the axillary bud, resulting in a new in vitro plantlet. This type of propagation promotes the development of a pre-existent morphological structure. The nutritional and hormonal condition of the medium breaks the dormancy of the axillary bud and promotes its rapid development.

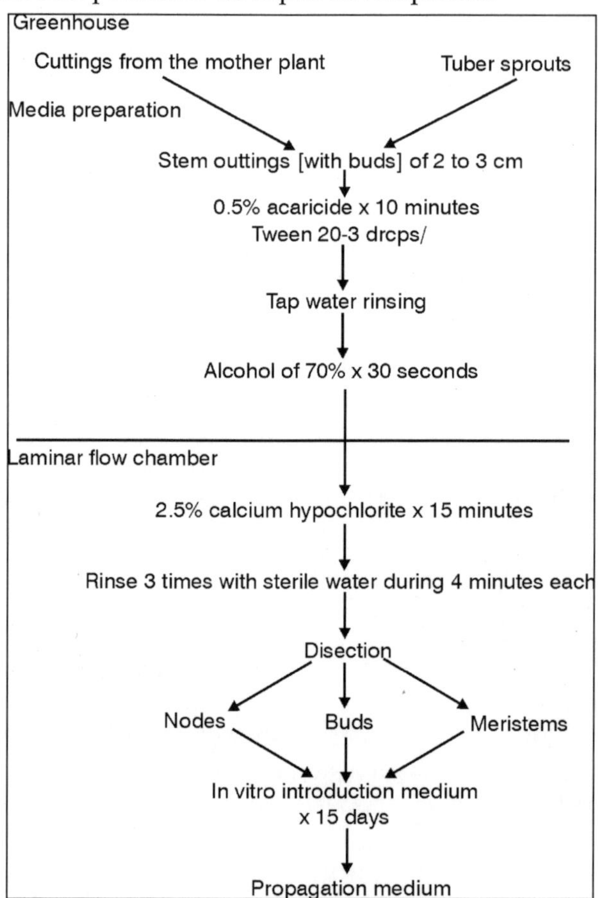

Fig. Procedure for in vitro Introducction

Callus formation and plant regeneration must be avoided because they tend to affect the genetic stability of the genotype. Under room-controlled conditions micropropagation is fast. Each node planted in a propagation medium will produce a plantlet which will occupy the full length of the test tube, after approximately four weeks for potato, and six weeks for sweetpotato. The resultant in vitro plantlets may be transplanted to in vitro conditions in small pots in the greenhouse.

MICROPROPAGATION BY NODE CUTTINGS IN A LIQUID MEDIUM

This technique is applied both with potato and sweetpotato to produce a large number of nodes rapidly. Stem cuttings with 5 to 8 nodes are prepared by removing both the apex and the root of the in vitro plant to be propagated. The stems are placed in the corresponding propagation liquid medium. It is also possible to use isolated nodes: the nodes will sprout and new plantlets will develop over a period of 3 to 4 weeks.

Micropropagation Procedure

- Sterilize petri dishes (placed in paper bags or comets) and prepare the laminar flow chamber by disinfecting the internal surfaces with alcohol.

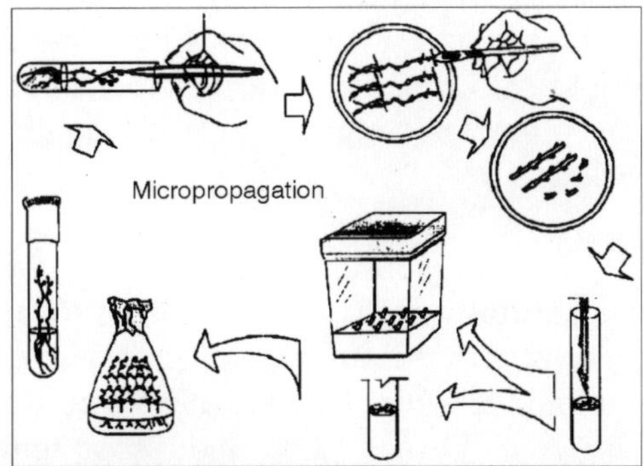

Fig. Potato Micropropagation Process Scheme

Sterilize the tools with an instrument sterilizer and place them on a sterile dish.

- Open the tube, take off the plantlet and place it on a petri dish with the help of forceps.
- Remove the leaves and cut the nodes.
- Open a tube containing fresh sterile medium and place a node inside, trying to plunge it slightly into the medium with the bud up. Close the tube.
- Seal the tube with a gas-permeable plastic tape (parafilm or saram wrap) and label it correctly. It is recommended to place two explants in 16 × 125 mm tubes, three in 18 × 150 mm tubes, five in 25 × 150 mm tubes, and 20-30 in magenta vessels.

COMMON PROBLEMS IN MICROPROPAGATION

Some problems may appear in tissue culture according to the crop or variety. To solve them, it is necessary to apply one or several preventing/ solving methods such as:

Phenolization

The explants frequently become brown or blackish shortly after isolation. When this occurs, growth is inhibited and the tissue generally dies. The young tissues are less susceptible to darkening than the more mature ones.

Prevention

The tissue darkening– mainly that of the recently isolated explants and that of the medium– may be generally prevented by:

- Removing the phenolic compounds produced by dispersion. Absorption by means of activated carbon. Absorption by polyvinilpyrolidone (PVP).
- Modifying the redox potential. Reducing agents: ascorbic acid, citric acid, L-cisteine HCL, ditriotreitol, glutation and mercaptoethanol Less availability of oxygen: stationary liquid media.
- Inactivating the phenolase enzymes. Chelating agents: NaFeEDTA, EDTA, diethyldithiocarbamate, dimethyl-dithiocarbamate
- Reducing the phenolasic activity and the availability of substrate. Low pH Darkness

Absence of Rooting

The explants can naturally form roots during propagation, without an additional rooting stage, as with the potato. However, some wild potato species may show root production deficiency. Rooting may be induced by incorporating auxins, such as IAA, NAA, and IBA, or activated carbon to the culture medium.

POTATO IN VITRO TUBERIZATION

Most of the potato microtubers are used in seed programmes in Europe, where large amounts (hundred of thousands) of pre-basic seed of a few varieties are produced. By means of this technique, microtubers are produced and stored, and it is possible to store thousands of them in a small area (in humid containers at 4°C) for long periods of time. The microtuber induction is produced through a stress effect by the CCC (chlorocholine chloride), BAP (B-benzylaminopurine) and Sucrose, which under darkness will produce from 3 to 4 microtubers per plant, according to the variety.

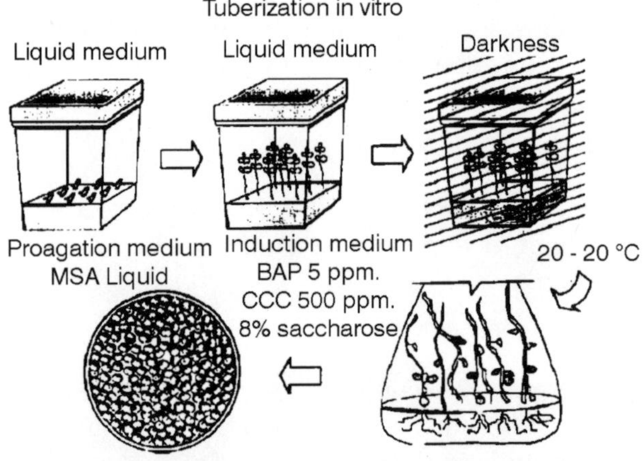

Fig. Poato in Vitro Tuberization Process Scheme

In the beginning, microtubers were used as an alternative for germplasm distribution and in vitro conservation. Curing the plantlet in vitro multiplication process, usually there is a multiplication of higher amounts to those required in the greenhouse and so, after the transference, some magentas are left over. These plantlets may be used for microtuber induction.

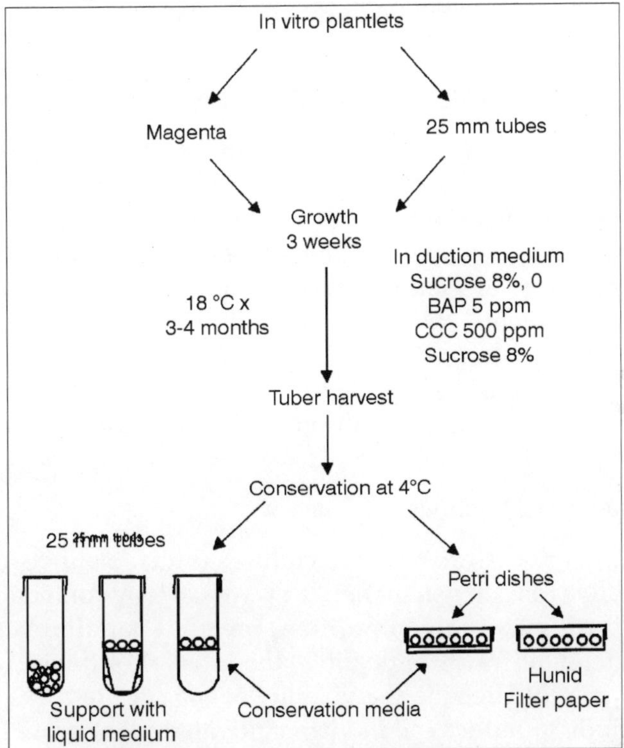

Fig. Potato in Vitro Tuberization Process Scheme

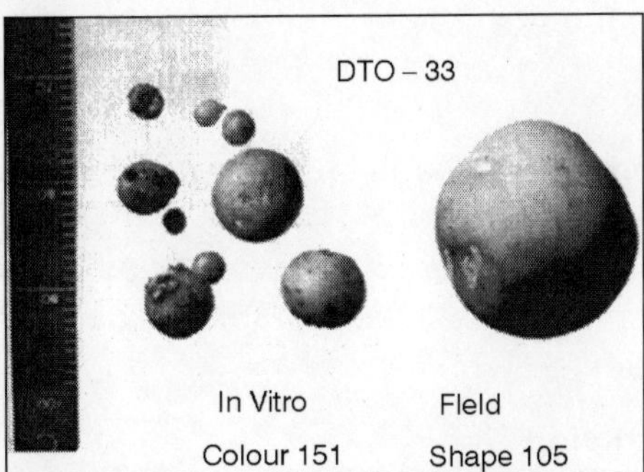

Fig. Comparison, in Size and Shape of in Vitro Microtubers with a Normal one from the Field

In accordance with the indicated procedures, the induction medium is added to the magentas and then they are placed in the dark room. After three months the microtubers will be produced. These may be harvested and transferred to sterile containers (at 4°C), where they can be maintained up to 10 months, or used in the next campaign instead of in vitro plantlets.

These microtubers can also be used as a reserve, in case in vitro material is contaminated or dies, because of temperature or handling effects. The high number and bigger size of the microtubers increases the success rates for the transfer to the greenhouse.

MEDIUM TERM GERMPLASM MAINTENANCE

The in vitro germplasm maintenance under normal growth conditions requires a series of transfers of the plantlets into a fresh medium. This leads to a consumption of time, increases the possibility of losses because of material contamination during successive sub-cultures, causes a loss of material by human error or failure of some equipment, and demands more labour A way to avoid these problems is through limitation, restriction or inhibition of growth.

This approach consists of growth speed reduction by modifying the physical or chemical conditions of the culture, and it is effective for a short or medium term period.

In Vitro Conservation Methods

These consist in maintaining the cultures (buds, plantlets derived from nodes or directly from meristems) under physical (environmental factors] or chemical (culture medium composition) stressed conditions that make it possible to extend, as much as possible, the interval of transference into the fresh media, without affecting the viability of the cultures.

The methods to reduce the in vitro cultivated plant growth include the reduction of temperature and light during storage, the incorporation of growth retardants in the medium, and the induction of osmotic stress in the medium, or a combination of all these.

Temperature

Temperature reduction has been the most used way to curb culture growth.

Most of the in vitro cultures are maintained at temperatures between 12 and 200C: at lower temperatures, the growth rate decreases but the reduction depends on the species.

Nutrients Concentration

The reduction of the carbohydrate concentration and the nitrogenous components of the nutritive medium may affect the growth rates. In addition, the continuous absorption of these nutrients during plantet growth will bring a nutritional defficiency, which could produce a premature death of the plants.

Use of Growth Regulators

The abscisic acid (ABA), phosphon-D, maleic hydrazide, and succinic acid are some of the most frequently used growth regulators.

Osmotic Concentration

Growth limitation caused by osmotic concentration is due to the reduction of the water and nutrients absorbed from the medium. For example, at high concentrations, sucrose acts osmotically and it is highly metabolized. No metabolized osmotics, such as Manitol and Sorbitol, are possibly more efective than sucrose in culture growth limitation.

Evaluation of in Vitro Preserved Material

To evaluate material under these conditions, some important facts about in vitro maintenance such as genetic viability and stability should be considered. The viability evaluation of the in vitro cultures must be systematic. In slow growth conditions, when the sub-culture or transfer period extends during months or years, the frequency of the culture evaluation increases. The most important characteristics to be evaluated in the low-growth cultivars storage of apical buds are: contamination, leaf senescence, the number of green sprouts, the number of viable nodes in relation to the stem length, the presence or not of roots, and callus formation.

MAINTENANCE OF ACCESSIONS IN SEED PROGRAMME

In a seed programme, a group of accessions free of virus is maintained for pre-basic seed production: however, most of them are not propagated in the greenhouse so they are maintained in vitro to be used in the future. The continuous propagation of the in vitro plantlets damages the material, mainly if we consider that environmental conditions are not adequate.

The alternative is to maintain the accessions in conservation media. Each accession must be maintained in test tubes with five replications to avoid possible losses. The maintenance conditions have been tested in CIP's potato collection that consists of more than 5,000 accessions, by means of which it is possible to assure the normal recovery of the material after using stress producers.

The plantlets used in each campaign must come from the maintenance phase. Then, they will be propagated in normal media where their growth will be reestablished, and the elected procedures for plantlet multiplication in the greenhouse will continue.Each season must be initiated with this material to start off with strong plants. In addition, the plantlets taken for multiplication must be replaced, trying to maintain 5 tubes per accession in the conservation media.

The sub-cultures in the conservation medium are carried out approximately every one or two years. The conservation medium renovation must pass through a previous sub-culture in the propagation media to rejuvenate the explants.

VIRUS ERADICATION THROUGH MERISTEMS CULTURE AND THERMOTHERAPY

If a healthy plant is sown in the field, it is exposed to infections caused by pathogens as nematodes, fungi, bacteria, phytoplasms, virus and viroids, which have a negative effect on yield, and in some cases may kill the plants. However, not all the plant cells may become infected. A group of cells, which are in a continuous non-differentiated multiplication, are virus-free: the meristem. The in vitro meristems culture, together with growth at high temperatures produces potato plantlets free from viruses in more than 90 per cent of planted meristems. This routine method was established in the International Potato Centre to obtain virus-free plantlets for national and international distribution. The in vitro maintenance of virus-free plants provides the possibility to maintain all the time a bank of healthy and more vigourous plants, and with a more accelerated growth, than the infected ones. In a seed programme, it is essential to initiate this work with virus-free plantlets since this will affect the seed quality and the yield as well. The virus cleaning procedure is too long and expensive: that is why CIP, through its germplasm distribution programme, has provided a list of the principal virus-free potato varieties to all users.

PROCEDURE

- Approximately 18 to 20 plantlets (virus-infected) are propagated in magentas.
- After a growth period of 20 to 25 days, or when the plantlets are 4-5 cm high, they are placed in the thermotherapy chamber. The growth conditions are: 16 hours of light 34°C; 8 hours of darkness 32°C
- The magentas are maintained in the thermotherapy chamber for one month.
- Afterwards, the magentas are taken out of the chamber, and the outside is cleaned with 980/o alcohol. Then they are introduced to the culture room.
- Meristems are obtained as follows:

Cut the apical portion and remove the leaves that cover the meristem (approximately 3 to 4 leaves); the meristem is observed with a prominent leaf primordium. Remove the meristem with part of the leaf primordium; cut only the translucent portion. Use a new knife. Place the menistem in the culture medium. Be sure that the meristem is in the tube.

- Evaluate meristem growth and transfer them to fresh media if necessary.

- Each meristem that originates a plant is called <'line», which will be labeled according to the accession it belongs to. For example, Yellow Line 1, Yellow Line 2, Yellow Line 3, etc.
- Five tubes with several plants are propagated to evaluate the potato virus PSTVd, PVT, to determine the host range, the morphologic evaluation, and the in vitro maintenance.
- The results of each evaluation are obtained, and the infected material is replaced with clean material.

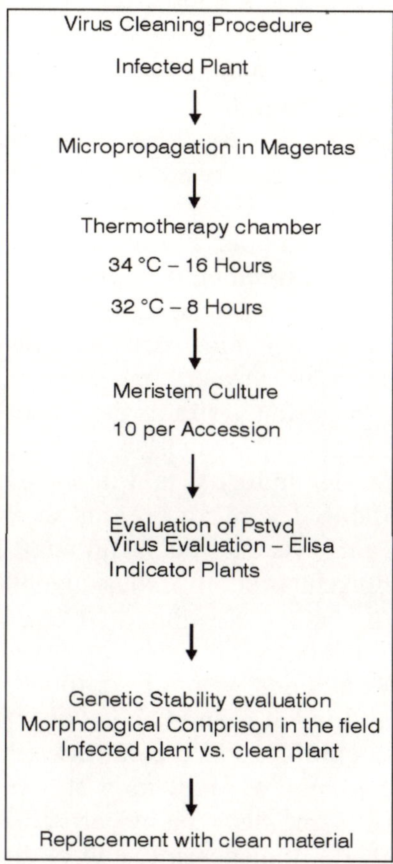

Temperature

The growth of the plantlets at temperatures higher than 20°C is accelerated (Do not use temperatures higher than 30°C).

CULTURED CELLS AND TISSUES

The large scale commercial propagation of plant material based on plant tissue culture was pioneered in the USA. During the last thirty years, tissue culture-based plant propagation has emerged as one of the leading global agro-technologies. Between 1986 and 1993, the worldwide production of tissue cultured plants increased 50 per cent.

In 1993, the production was 663 million plants. By 1997, production had risen to 800 million plants. During 1990–1994, the micropropagation industry declined in Europe, mainly due to production shifting to developing countries, but since then because of the demand for high quality and number, production in European countries has increased.

Since 1995, production has increased by 14 per cent in Asian countries, mainly due to the market entry of China, while the increase in South and Central America was from production in Cuba. More recently, some companies from Israel, the USA and UK have shifted their production requirements to Costa Rica and India.

Tissue cultured plants have as yet to reach many growers and farmers in the developing countries. The primary advantage of micropropagation is the rapid production of high quality, disease-free and uniform planting material. The plants can be multiplied under a controlled environment, anywhere, irrespective of the season and weather, on a year-round basis. Production of high quality and healthy planting material of ornamentals, and forest and fruit trees, propagated from vegetative parts, has created new opportunities in global trading for producers, farmers, and nursery owners, and for rural employment. Micropropagation technology is more expensive than the conventional methods of plant propagation, and requires several types of skills. It is a capital-intensive industry, and in some cases the unit cost per plant becomes unaffordable. The major reasons are cost of production and know-how. During the early years of the technology, there were difficulties in selling tissue culture products because the conventional planting material was much cheaper.

Now this problem has been addressed by inventing reliable and cost effective tissue culture methods without compromising on quality. This requires a constant monitoring of the input costs of chemicals, media, energy, labour and capital. In the industrialized countries, labour is the main factor that contributes to the high cost of production of tissue-cultured plants.

To reduce such costs, some steps can be partially mechanized, *e.g.* use of a peristaltic pump for medium dispensing, and of dishwashers for cleaning containers. In the less developed countries of Africa, Asia, and Latin America, where labour is relatively cheaper, consumables such as media, culture containers, and electricity make a comparatively greater contribution to production costs.

For example, the cost of medium preparation (chemicals, energy and labour) can account for 30–35 per cent of the micropropagated plant production. However, automated production processes based on pre-sterilized membrane capsules, bioreactors, mechanized explant transfer, and container sealing are not commercially viable propositions in many developing countries. Therefore, low cost alternatives are needed to reduce production cost of tissue-cultured plants.

Many of the low cost technology options described in this publication can be incorporated in various steps of plant micropropagation. The occasional tissue culturegenerated variants (somaclones) and rare spontaneous bud mutants as well as those obtained from induced mutations can also be propagated by deployment of low cost techniques described. Micropropagation, in combination with radiation-induced mutations, speeds up the recovery, multiplication and release of improved varieties in vegetatively propagated plants. Hence, low cost technology will be of great value for large scale plant multiplication of mutants of many fruits, shrubs, flowers and forest trees that are conventionally vegetatively propagated.

PLANT TISSUE CULTURE

Plant tissue culture refers to growing and multiplication of cells, tissues and organs of plants on defined solid or liquid media under aseptic and controlled environment. The commercial technology is primarily based on micropro-pagation, in which rapid proliferation is achieved from tiny stem cuttings, axillary buds, and to a limited extent from somatic embryos, cell clumps in suspension cultures and bioreactors. The cultured cells and tissue can take several pathways. The pathways that lead to the production of true-to-type plants in large numbers are the preferred ones for commercial multiplication. The process of micropropagation is usually divided into several stages *i.e.,* prepropagation, initiation of explants, subculture of explants for proliferation, shooting and rooting, and hardening. These stages are universally applicable in large-scale multiplication of plants. The delivery of hardened small micropropagated plants to growers and market also requires extra care.

Plant tissue culture refers to growing and multiplication of cells, tissues and organs on defined solid or liquid media under aseptic and controlled environment. Plant tissue culture technology is being widely used for large-scale plant multiplication. The commercial technology is primarily based on micropropagation, in which rapid proliferation is achieved from tiny stem cuttings, axillary buds, and to a limited extent from somatic embryos, cell clumps in suspension cultures and bioreactors.

EXPLANT SOURCE

Plant tissue cultures are initiated from tiny pieces, called explants, taken from any part of a plant. Practically all parts of a plant have been used successfully as a source of explants. In practice, the "explant" is removed surgically, surface sterilized and placed on a nutrient medium to initiate the mother culture, that is multiplied repeatedly by subculture. The following plant parts are extensively used in commercial micropropagation. Shoot-tip and meristem-tip culture: Shoots develop from a small group of cells known as shoot apical meristem. The apical meristem maintains itself, gives rise to new tissues and organs, and communicates signals to the rest of the plant. Shoot-tips and meristem-tips are perhaps the most popular source of explants to initiate tissue cultures. The shoot apex explant measures between 100 to 500 μ m and includes the apical meristem with 1 to 3 leaf primordia. The apical meristem of a shoot is the portion lying distal to the youngest leaf primordium, and is ca.100 μ??m in diameter and 250μ?m in length with 800-1200 cells. In practice, shoot-tip explants between 100 to 1000 μ m are cultured to free plants from viruses. Even explants larger than 1000 μ m have been frequently used. The term "meristem-tip culture" has been suggested to distinguish the large explants from those used in conventional propagation.

Nodal or axillary bud culture: This consists of a piece of stem with axillary bud culture with or without a portion of shoot. When only the axillary bud is taken, it is designated as "axillary bud" culture. Floral meristem and bud culture: Such explants are not commonly used in commercial propagation, but floral meristems and buds can generate complete plants. Other sources of explants: In some plants, leaf discs, intercalary meristems from nodes, small pieces of stems, immature zygotic embryos and nucellus have also been used as explants to initiate cultures.

Cell suspension and callus cultures: Plant parts such as leaf discs, intercalary meristems,-stem-pieces, immature embryos, anthers, pollen, microspores and ovules have been cultured to initiate callus. A callus is a mass of unorganized cells, which in many cases, upon transfer to suitable medium, is capable of giving rise to shoot-buds and somatic embryos, which then form complete plants. Such calli on culture in liquid media on shakers are used for initiating cell suspensions. Liquid suspension cultures maintained on mechanical shakers achieve fast and excellent multiplication rates. However, in commercial micropropagation, calli are cultured mostly in bottles and flasks kept on semi-solid or liquid media. To a limited extent, bioreactors have become popular for somatic embryogenic cultures. It is considered that some day robotics could be adapted to bioreactorbased micropropagation.

PATHWAYS OF CULTURED CELLS AND TISSUES

The cultured cells and tissue can take several pathways to produce a complete plant. Among these, the pathways that lead to the production of true-to-type plants in large numbers are the popular and preferred ones for commercial multiplication. The following terms have been used to describe various pathways of cells and tissue in culture.

Regeneration and Organogenesis

In this pathway, groups of cells of the apical meristem in the shoot apex, axillary buds, root tips, and floral buds are stimulated to differentiate and grow into shoots and ultimately into complete plants. In many cases, the axillary buds formed in the culture undergo repetitive proliferation, and produce large number of tiny plants.

The plants are then separated from each other and rooted either in the next stages of micropropagation or in vivo (in trays, small pots or beds in glasshouse or plastic tunnel under relatively high humidity). The explants cultured on relatively high amounts of auxin (*e.g.* (2,4-D, 2,4-dichlorophenoxyacetic acid) form an unorganized mass of cells, called callus. The callus can be further sub-cultured and multiplied.

The callus shaken in a liquid medium produces cell suspension, which can be subcultured and multiplied into more liquid cultures. The cell suspensions form cell clumps, which eventually form calli and give rise to plants through organogenesis or somatic embryogenesis. In some cases, explants *e.g.* leaf-discs and epidermal tissue can also generate plants by direct organogenesis and somatic embryogenesis without intervening callus formation, *e.g.* in orchardgrass. Dactylis glomerata L.. In organogenesis the cultured plant cells and cell clumps (callus) and mature differentiated cells (microspores, ovules) and tissues (leaf discs, inter-nodal segments) are induced to differentiate into complete plants to form shoot buds and eventually shoots, and rooted to form complete plants.

Somatic Embryogenesis

In this pathway, cells or callus cultures on solid media or in suspension cultures form embryo-like structures called somatic embryos, which on germination produce complete plants. The primary somatic embryos are also capable of producing more embryos through secondary somatic embryogenesis. Although, somatic embryogenesis has been demonstrated in a very large number of plants and trees, the use of somatic embryos in large-scale commercial production has been restricted to only a few plants, such as carrot, date palm, and a few forest trees.

Somatic embryos are produced as adventitious structures directly on explants of zygotic embryos, from callus and suspension cultures. Somatic embryos and synthetic seeds (embryos encapsulated in artificial endosperm) hold potential for large-scale clonal propagation of superior genotypes of heterogeneous plants.

They have also been used in commercial plant production and for the multiplication of parental genotypes in large-scale hybrid seed production. In many species, somatic embryos are morphologically similar to the zygotic embryos, although some biochemical, physiological and anatomical differences have been documented.

The synthetic auxin, 2,4-D is commonly used for embryo induction. In many angiosperms, *e.g.*, carrot and alfalfa, subculture of cells from 2,4-D containing medium to auxin-free medium is sufficient to induce somatic embryogenesis. The process can be enhanced with the application of osmotic stress, manipulation of medium nutrients, and reducing humidity. Selection of embryogenic cell lines has also been successfully used. For example, selection for unique morphotypes in grapevine cultures allows production of high quality embryos with predicable frequency.

A major problem in large-scale production of somatic embryos is culture synchronization. This is achieved through selecting cells or pre-embryonic cell clusters of certain size, and manipulation of light and temperature, temporary starvation or by adding cell cycle synchronizing chemicals to the medium. Cytokinins seem to play a key role in cell cycle synchronization and embryo induction, proliferation and differentiation. Abscisic acid is crucial in all the stages of somatic development, maturation and hardening.

Synthetic Seeds

The concept of production and utilization of synthetic seeds (somatic embryo as substitutes for true seeds) was first suggested by Murashige in 1977. Synthetic seeds can be produced either as coated or non-coated, desiccated somatic embryos or as embryos encapsulated in hydrated gel (usually calcium alginate).

Successful utilization of synthetic seeds as propagules of choice requires an efficient and reproducible production system and a high percentage of post-planting conversion into vigourous plants. Artificial coats and gel capsules containing nutrients, pesticides and beneficial organisms have long been thought as substitutes for seed coat and endosperm. However, this technology is still in the developmental stage, and currently cannot compete with the other methods of commercial plant propagation.

Process of Micropropagation

The process of plant micro-propagation aims to produce clones (true copies of a plant in large numbers).

The process is usually divided into the following stages:

Stage 0 – Pre-propagation step or selection and pre-treatment of suitable plants.

Stage I – Initiation of explants-surface sterilization, establishment of mother explants.

Stage II – subculture for multiplication/proliferation of explants.

Stage III – shooting and rooting of the explants.

Stage IV – Weaning/hardening.

These stages are universally applicable in large-scale multiplication of plants. The individual plant species, varieties and clones require specific modification of the growth media, weaning and hardening conditions.

A rule of the thumb is to propagate plants under conditions as natural or similar to those in which the plants will be ultimately grown ex-vitro. For example, if a chrysanthemum variety is to be grown under long day-length for flower production, it is better to multiply the material under long-day length at stages III and IV. There is a wide option to undertake production of plant material up to a limited number of stages. For example, many commercial tissue culture companies undertake production up to Stage III, and leave the remaining stages to others.

Pre-Propagation Stage

The pre-propagation stage (also called stage 0) requires proper maintenance of the mother plants in the greenhouse under disease- and insect-free conditions with minimal dust. Clean enclosed areas, glasshouses, plastic tunnels, and net-covered tunnels, provide high quality explant source plants with minimal infection. Collection of plant material for clonal propagation should be done after appropriate pretreatment of the mother plants with fungicides and pesticides to minimize contamination in the in vitro cultures. This improves growth and multiplication rates of in vitro cultures.

The control of contamination begins with the pretreatment of the donor plants. They may be prescreened for diseases, isolated and treated to reduce contamination. The explants are then brought to the production facility, surface sterilized and introduced into culture. They may at this stage be treated with antibiotics and fungicides as well as anti-microbial formulations, such as PPM. The explants are then culture indexed for contamination by standard microbiological techniques, which are occasionally supplemented with tests based on molecular biology or other techniques.

Stage I

This stage refers to the inoculation of the explants on sterile medium to initiate aseptic culture. Initiation of explants is the very first step in micropropagation. A good clean explant, once established in an aseptic condition, can be multiplied several times; hence, explant initiation in an aseptic condition should be regarded as a critical step in micropropagation. More than often, explants fail to establish and grow, not due to the lack of a suitable medium but because of contamination. The explants are transferred to in vitro environment, free from microbial contaminants. The process requires excision of tiny plant pieces and their surface sterilization with chemicals such as sodium hypochlorite, ethyl alcohol and repeated washing with sterile distilled water before and after treatment with chemicals. After a short period of culture, usually 3 to 5 days, the contaminated explants are discarded.

The surviving explants showing growth are maintained and used for further subculture. In herbaceous plants *e.g.* potato, chrysanthemum, carnation, streptocarpus, strawberry, and African violet; the explant sources are meristems, apical- and axillary buds, young seedlings, developing young leaves and petioles, and unopened floral buds. The following low cost options can be adapted to initiate explants:

Sterile Instrument Technique

This method assumes that most of the deep-seated meristems and those covered by leaves or other integuments (*e.g.* floral bracts) are sterile. In this procedure, the explant is washed with sterile water, rinsed in ethanol, and instruments are sterilized every time they touch the surface of the explant, and the explant is moved to a new location on the dissection stage.

Surface Sterilization Technique

This is by far the most commonly used method. The explants are washed in sterile water, rinsed in ethanol, and surface sterilization is achieved by using chemicals with chlorine base. Calcium or sodium hypochlorite based solutions, 1–3 per cent (v/v) are usually used for soft herbaceous materials. A cheap and ready-made sterilant is 5-7 per cent solution of 'Domestos'- a toilet disinfectant which contains 10.5 per cent v/v sodium hypochlorite, 0.3 per cent sodium carbonate, 10.0 per cent sodium chloride and 0.5 per cent (w/v) sodium hydroxide and a patented thickener). The explants are washed in sterile distilled water before and after sterilization. Other surface sterilants used include mercuric chloride (avoid its use as far as possible, since it is highly toxic), hydrogen peroxide, and potassium permanganate.

For Soft Tissues

- Wash explants from perennial plants for 1-2 hr in tap water. Eliminate this step for material from glasshouse grown plants.
- Wash in sterile distilled water three to four times for 5 to 10 minutes each.
- Dip in 95 per cent ethanol for 3 to 5 seconds.
- Wash once again with sterile distilled water for 5 minutes.
- Surface-sterilize in 5 per cent 'Domestos' (v/v) for 20-25 minutes.
- Wash with sterile distilled water three times for 10 minutes each.
- Drain water droplets by placing on pre-sterilized blotting paper.
- Transfer explants singly to the medium.

N.B.: Sterilize forceps each time to transfer explants to avoid cross-contamination.

For woody stems (e.g. roses, hardy shrubs, and trees):

- Collect stems, shoots, buds and store at 5 oC till needed.
- Rinse in ethanol for 3 to 5 seconds.
- Rinse in 1- per cent sodium hypochlorite (20 per cent bleach) for 10 minutes.
- Place lower parts of stems in flasks in 2 per cent sucrose and 200 PPM 8-hydroxyquinoline citrate at 23+2 oC. For items collected in September/October, add 50-PPM GA3.

After that 10 PPM GA will help break the dormancy:

- Re-cut the bottom of stem and replace the solution after 2 days.
- Excise the softwood from the developed shoots and use material for explants or for rooting.
- Surface-sterilize as in the above protocol.

Do not forget to sterilize forceps and scalpel every time for the transfer of explants to fresh solutions. Use sterile containers in the protocol of surface sterilization. If explants become brown or pale at the end of the protocol, reduce the strength of 'Domestos' to 2.5 per cent. Alternatively, dip explants in 10 per cent 'Domestos' for 2 minutes and then proceed to surface sterilize with 3-5 per cent 'Domestos' for 20 minutes. If basal contamination is observed after 2-3 days of culture, explants can sometimes be rescued by removing the basal end by making a single cut with a sharp scalpel and re-culturing on fresh medium.

Stage II

Stage II is the propagation phase in which the explants are cultured on the appropriate media for multiplication of shoots. The primary goal is to achieve propagation without losing the genetic stability. Repeated culture of axillary and adventitious shoots, cutting with nodes, somatic embryos and other organs from Stage I leads to multiplication of propagules in large numbers. The propagules produced at this stage can be further used for multiplication by their repeated culture. Sometimes it is necessary to subculture the in vitro derived shoots onto different media for elongation.

Stage III

The in vitro shoots obtained at Stage II are rooted to produce complete plants. If the proliferated material consists of bud-like structures (*e.g.* orchids) or clumps of shoots (banana, pineapple), they should be separated after rooting and not before. Many plants (*e.g.* banana, pineapple, roses, potato, chrysanthemum, strawberry, mint, several grasses and many more) can be rooted on half-strength-MS medium without any growth-regulators. Good sturdy well-rooted plants are essential for high survival during weaning and later transfer to soil. This stage is labour intensive and expensive. The process of in vitro rooting has been estimated to account for approximately 35–75 per cent of the total cost of production. Efforts should be made to combine rooting and acclimatization stages.

Stage IV

At this stage, the in vitro micropropagated plants are weaned and hardened. This is the final stage of the tissue culture operation after which the micropropagated plantlets are ready for transfer to the greenhouse. Steps are taken to grow individual plantlets capable of carrying out photosynthesis.

The hardening of the tissue-cultured plantlets is done gradually from high to low humidity and from low light intensity to high intensity conditions. If grown on solid medium, most of the agar can be removed gently by rinsing with water. Plants can be left in shade for 3 to 6 days where diffused natural light conditions them to the new environment. The plants are then transferred to an appropriate substrate (sand, peat, compost, etc.), and gradually hardened. Low-cost options include the use of plastic domes or tunnels, which reduces the natural light intensity and maintains high relative humidity during the hardening process. If the plants are still joined together after rooting, these should be planted as bunches in the soil and separated after 6 to 8 weeks of growth.

DELIVERY TO THE GROWERS

The delivery of the rooted and hardened small micropropagated plants to growers and market requires extra care. In some cases, plant losses can occur during shipment and handling by growers. This is particularly true when the plants are not fully hardened and rooted or not grown for sufficient duration after transfer to soil. Growers should be given clear instructions how to handle the material provided. Apart from the economic loss, poor survival of planted material erodes the confidence of growers in the technology. The transfer of individual plants to soil in black plastic or polythene bags is widely used as a low-cost option to provide fully-grown banana plants directly to farmers in many developing countries.

PLANT GROWTH REGULATORS

We have already briefly considered the concepts of plasticity and totipotency. The essential point as far as plant cell culture is concerned is that, due to this plasticity and totipotency, specific media manipulations can be used to direct the development of plant cells in culture.

Plant growth regulators are the critical media components in determining the developmental pathway of the plant cells. The plant growth regulators used most commonly are plant hormones or their synthetic analogues.

Classes of Plant Growth Regulators

There are five main classes of plant growth regulator used in plant cell culture, namely:

(1) auxins;
(2) cytokinins;
(3) gibberellins;
(4) abscisic acid;
(5) ethylene.

Each class of plant growth regulator will be briefly looked at.

Auxins

Auxins promote both cell division and cell growth The most important naturally occurring auxin is IAA (indole-3-acetic acid), but its use in plant cell culture media is limited because it is unstable to both heat and light. Occasionally, amino acid conjugates of IAA (such as indole-acetyl-L-alanine and indole-acetyl-L-glycine), which are more stable, are used to partially alleviate the problems associated with the use of IAA. It is more common, though, to use stable chemical analogues of IAA as a source of auxin in plant cell culture media. 2,4-Dichlorophenoxyacetic acid (2,4-D) is the most commonly used auxin and is extremely effective in most circumstances. Other auxins are available, and some may be more effective or 'potent' than 2,4-D in some instances.

CYTOKININS

Cytokinins promote cell division. Naturally occurring cytokinins are a large group of structurally related (they are purine derivatives) compounds. Of the naturally occurring cytokinins, two have some use in plant tissue culture media. These are zeatin and 2iP (2-isopentyl adenine). Their use is not widespread as they are expensive (particularly zeatin) and relatively unstable. The synthetic analogues, kinetin and BAP (benzylaminopurine), are therefore used more frequently. Non-purine-based chemicals, such as substituted phenylureas, are also used as cytokinins in plant cell culture media. These substituted phenylureas can also substitute for auxin in some culture systems.

GIBBERELLINS

There are numerous, naturally occurring, structurally related compounds termed 'gibberellins'. They are involved in regulating cell elongation, and are agronomically important in determining plant height and fruit-set. Only a few of the gibberellins are used in plant tissue culture media, Gibberelic Acid 3 (GA3) being the most common.

Abscisic Acid

Abscisic acid (ABA) inhibits cell division. It is most commonly used in plant tissue culture to promote distinct developmental pathways such as somatic embryogenesis.

Ethylene

Ethylene is a gaseous, naturally occurring, plant growth regulator most commonly associated with controlling fruit ripening, and its use in plant tissue culture is not widespread. It does, though, present a particular problem for plant tissue culture. Some plant cell cultures produce ethylene, which, if it builds up sufficiently, can inhibit the growth and development of the culture. The type of culture vessel used and its means of closure affect the gaseous exchange between the culture vessel and the outside atmosphere and thus the levels of ethylene present in the culture.

Plant Growth Regulators and Tissue Culture

Generalisations about plant growth regulators and their use in plant cell culture media have been developed from initial observations made in the 1950s. There is, however, some considerable difficulty in predicting the effects of plant growth regulators: this is because of the great differences in culture response between species, cultivars and even plants of the same cultivar grown under different conditions.

However, some principles do hold true and have become the paradigm on which most plant tissue culture regimes are based. Auxins and cytokinins are the most widely used plant growth regulators in plant tissue culture and are usually used together, the ratio of the auxin to the cytokinin determining the type of culture established or regenerated. A high auxin to cytokinin ratio generally favours root formation, whereas a high cytokinin to auxin ratio favours shoot formation. An intermediate ratio favours callus production.

CELL-SUSPENSION CULTURES

Callus cultures, broadly speaking, fall into one of two categories: compact or friable. In compact callus the cells are densely aggregated, whereas in friable callus the cells are only loosely associated with each other and the callus becomes soft and breaks apart easily. Friable callus provides the inoculum to form cell-suspension cultures. Explants from some plant species or particular cell types tend not to form friable callus, making cell-suspension initiation a difficult task.

The friability of callus can sometimes be improved by manipulating the medium components or by repeated subculturing. The friability of the callus can also sometimes be improved by culturing it on 'semi-solid' medium (medium with a low concentration of gelling agent). When friable callus is placed into a liquid medium (usually the same composition as the solid medium used for the callus culture) and then agitated, single cells and/or small clumps of cells are released into the medium.

Under the correct conditions, these released cells continue to grow and divide, eventually producing a cell-suspension culture. A relatively large inoculum should be used when initiating cell suspensions so that the released cell numbers build up quickly. The inoculum should not be too large though, as toxic products released from damaged or stressed cells can build up to lethal levels. Large cell clumps can be removed during subculture of the cell suspension. Cell suspensions can be maintained relatively simply as batch cultures in conical flasks. They are continually cultured by repeated subculturing into fresh medium. This results in dilution of the suspension and the initiation of another batch growth cycle. The degree of dilution during subculture should be determined empirically for each culture. Too great a degree of dilution will result in a greatly extended lag period or, in extreme cases, death of the transferred cells.

After subculture, the cells divide and the biomass of the culture increases in a characteristic fashion, until nutrients in the medium are exhausted and/or toxic by-products build up to inhibitory levels—this is called the 'stationary phase'. If cells are left in the stationary phase for too long, they will die and the culture will be lost. Therefore, cells should be transferred as they enter the stationary phase. It is therefore important that the batch growth-cycle parameters are determined for each cell-suspension culture.

PROTOPLASTS

Protoplasts are plant cells with the cell wall removed. Protoplasts are most commonly isolated from either leaf mesophyll cells or cell suspensions, although other sources can be used to advantage. Two general approaches to removing the cell wall (a difficult task without damaging the protoplast) can be taken—mechanical or enzymatic isolation.

Mechanical isolation, although possible, often results in low yields, poor quality and poor performance in culture due to substances released from damaged cells.

Enzymatic isolation is usually carried out in a simple salt solution with a high osmoticum, plus the cell wall degrading enzymes. It is usual to use a mix of both cellulase and pectinase enzymes, which must be of high quality and purity.

Protoplasts are fragile and easily damaged, and therefore must be cultured carefully. Liquid medium is not agitated and a high osmotic potential is maintained, at least in the initial stages. The liquid medium must be shallow enough to allow aeration in the absence of agitation. Protoplasts can be plated out on to solid medium and callus produced. Whole plants can be regenerated by organogenesis or somatic embryogenesis from this callus. Protoplasts are ideal targets for transformation by a variety of means.

ROOT CULTURES

Root cultures can be established *in vitro* from explants of the root tip of either primary or lateral roots and can be cultured on fairly simple media. The growth of roots *in vitro* is potentially unlimited, as roots are indeterminate organs. Although the establishment of root cultures was one of the first achievements of modern plant tissue culture, they are not widely used in plant transformation studies.

Shoot Tip and Meristem Culture

The tips of shoots (which contain the shoot apical meristem) can be cultured *in vitro,* producing clumps of shoots from either axillary or adventitious buds. This method can be used for clonal propagation. Shoot meristem cultures are potential alternatives to the more commonly used methods for cereal regeneration as they are less genotype-dependent and more efficient (seedlings can be used as donor material).

Embryo Culture

Embryos can be used as explants to generate callus cultures or somatic embryos. Both immature and mature embryos can be used as explants. Immature, embryo-derived embryogenic callus is the most popular method of monocot plant regeneration.

Microspore Culture

Haploid tissue can be cultured *in vitro* by using pollen or anthers as an explant. Pollen contains the male gametophyte, which is termed the 'microspore'. Both callus and embryos can be produced from pollen. Two main approaches can be taken to produce *in vitro* cultures from haploid tissue.

The first method depends on using the anther as the explant. Anthers (somatic tissue that surrounds and contains the pollen) can be cultured on solid medium (agar should not be used to solidify the medium as it containsinhibitory substances). Pollen-derived embryos are subsequently produced via dehiscence of the mature anthers. The dehiscence of the anther depends both on its isolation at the correct stage and on the correct culture conditions. In some species, the reliance on natural dehiscence can be circumvented by cutting the wall of the anther, although this does, of course, take a considerable amount of time. Anthers can also be cultured in liquid medium, and pollen released from the anthers can be induced to form embryos, although the effi-ciency of plant regeneration is often very low. Immature pollen can also be extracted from developing anthers and cultured directly, although this is a very time-consuming process.

Both methods have advantages and disadvantages. Some beneficial effects to the culture are observed when anthers are used as the explant material. There is, however, the danger that some of the embryos produced from anther culture will originate from the somatic anther tissue rather than the haploid microspore cells.

If isolated pollen is used there is no danger of mixed embryo formation, but the efficiency is low and the process is time-consuming. In microspore culture, the condition of the donor plant is of critical importance, as is the timing of isolation. Pretreatments, such as a cold treatment, are often found to increase the efficiency. These pretreatments can be applied before culture, or, in some species, after placing the anthers in culture.

Plant species can be divided into two groups, depending on whether they require the addition of plant growth regulators to the medium for pollen/anther culture; those that do also often require organic supplements, *e.g.* amino acids. Many of the cereals (rice, wheat, barley and maize) require medium supplemented with plant growth regulators for pollen/anther culture.

Regeneration from microspore explants can be obtained by direct embryogenesis, or via a callus stage and subsequent embryogenesis.

Haploid tissue cultures can also be initiated from the female gametophyte (the ovule). In some cases, this is a more efficient method than using pollen or anthers.

The ploidy of the plants obtained from haploid cultures may not be haploid. This can be a consequence of chromosome doubling during the culture period. Chromosome doubling (which often has to be induced by treatment with chemicals such as colchicine) may be an advantage, as in many cases haploid plants are not the desired outcome of regeneration from haploid tissues. Such plants are often referred to as 'di-haploids', because they contain two copies of the same haploid genome.

CULTURE TYPES

Cultures are generally initiated from sterile pieces of a whole plant. These pieces are termed 'explants', and may consist of pieces of organs, such as leaves or roots, or may be specific cell types, such as pollen or endosperm. Many features of the explant are known to affect the efficiency of culture initiation. Generally, younger, more rapidly growing tissue (or tissue at an early stage of development) is most effective. Several different culture types most commonly used in plant transformation studies will now be examined in bit detail.

CALLUS

Explants, when cultured on the appropriate medium, usually with both an auxin and a cytokinin, can give rise to an unorganised, growing and dividing mass of cells. It is thought that any plant tissue can be used as an explant, if the correct conditions are found. In culture, this proliferation can be maintained more or less indefinitely, provided that the callus is subcultured on to fresh medium periodically.

During callus formation there is some degree of dedifferentiation (*i.e.* the changes that occur during development and specialization are, to some extent, reversed), both in morphology (callus is usually composed of unspecialised parenchyma cells) and metabolism. One major consequence of this dedifferentiation is that most plant cultures lose the ability to photosynthesise. This has important consequences for the culture of callus tissue, as the metabolic profile will probably not match that of the donor plant. This necessitates the addition of other components—such as vitamins and, most importantly, a carbon source—to the culture medium, in addition to the usual mineral nutrients.

Callus culture is often performed in the dark (the lack of photosynthetic capability being no drawback) as light can encourage differentiation of the callus. During long-term culture, the culture may lose the requirement for auxin and/or cytokinin. This process, known as 'habituation', is common in callus cultures from some plant species (such as sugar beet). Callus cultures are extremely important in plant biotechnology. Manipulation of the auxin to cytokinin ratio in the medium can lead to the development of shoots, roots or somatic embryos from which whole plants can subsequently be produced. Callus cultures can also be used to initiate cell suspensions, which are used in a variety of ways in plant transformation studies.

BANANA CELL AND TISSUE CULTURE

The International *Musa* germplasm collection is sited at the INIBAP (International Network for the Improvement of Banana and Plantain) Transit Centre at K.U.Leuven. By now, more than 1000 different accessions of shoot-tip cultures have been initiated *in vitro,* multiplied and maintained at reduced temperature conditions (16 ± 1°C). Shoot cultures are grown on MS (Murashige and Skoog) medium, supplemented with 30 g/l sucrose, 2.25 mg/l BA (6-benzyladenine) and 0.175 mg/l IAA (indole-3-acetic acid). In comparison with the culture medium on which shoot-tips are maintained, a tenfold decrease in cytokinin content (0.225 mg/l BA) induces regeneration of rooted plants. In contrast, adding 22.5 mg/l BA to the culture medium results in suppression of the apical dominance in shoot-tip cultures and a reduction of corm and leaf tissue between meristematic tissue.

Highly proliferating meristem cultures are obtained and used as starting material in the scalp methodology, the technique most commonly applied for the development of embryogenic cell suspensions at the Laboratory of Tropical Crop Improvement, K.U.Leuven. Initiation and maintenance of cell cultures is rather labour intensive and time consuming. However, since 1 ml of settled cells of a highly regenerable cell suspension can yield more than 100,000 plants, cell cultures are most suitable for mass clonal propagation. Moreover, embryogenic cell suspensions are highly preferred as target material for protoplast culture and genetic engineering since the risk of chimerism is circumvented because of the unicellular origin of regenerated plants.

Plant tissue culture is the science of growing plant cells, tissues or organs isolated from the mother plant, on artificial media. It includes techniques and methods appropriate to research into many botanical disciplines and several practical objectives. Both organized and unorganized growth is possible *in vitro*.

Organized growth of banana tissue *in vitro* is limited to embryo culture and shoot tip culture. Embryo culture is an important aid for classical breeding in banana since the germination frequency of seed is extremely low. Through embryo rescue, this frequency can be 10 times increased.

Major applications of shoot tip culture are mass clonal propagation and germplasm conservation. In the former already existing shoot tips are stimulated to multiply rapidly, while in the latter, the multiplication rate is slowed down. This technique is applied for the conservation of The International *Musa* germplasm collection at the INIBAP (International Network for the Improvement of Banana and Plantain) Transit Centre at K.U. Leuven.

In vitro culture of unorganized tissue in banana is almost exclusively related to the establishment of embryogenic cell cultures. When an embryogenic callus is induced on solid media containing high auxin concentrations, it can be transferred to liquid medium where it gives rise to embryogenic cell suspensions. Such suspensions with high regeneration capacity can be used for mass clonal propagation and are the only source of regenerable protoplasts in banana. More importantly, they are the preferred target material for induced mutations and genetic engineering.

In this chapter, a brief overview is given on the most common techniques in banana cell and tissue culture and their applications.

SHOOT-TIP CULTURES

Micropropagation

Stage 1: Initiation of shoot cultures

Shoot cultures of banana start conventionally from any plant part that contains a shoot meristem, *i.e.* the parental pseudostem, small suckers, peepers and lateral buds. The apex of the inflorescence and axillary flower buds are also suitable explants for tissue culture initiation. Overall, it is important to select explant material from preferably mature individuals whose response to environmental factors is known, and whose quality traits governed by genotypic and environmental effects have been identified.

For rapid *in vitro* multiplication of banana, shoot tips from young suckers of 40-100 cm height are most commonly used as explants. From the selected sucker a cube of tissue of about 1-2 cm^3 containing the apical meristem is excised. This block of tissue is dipped in 70% ethanol for 10 s, surface sterilized in a 2% sodium hypochlorite solution, and after 20 min rinsed three times for 10 min in sterile water. Variants of this decontamination protocol exist. They differ in explant type and size, disinfection procedure (single or double sterilisation), type of disinfectant (calcium hypochlorite instead of sodium hypochlorite) and its concentration and treatment duration. Subsequently a shoot tip of about 3 × 5 mm, consisting of the apical dome covered with several leaf primordia and a thin layer of corm tissue, is aseptically dissected. Larger explants have the merit of consisting of a shoot apex bearing more lateral buds which rapidly develop into shoots.

The optimal size of the explant depends on the purpose. For rapid multiplication, a relatively larger explant (3-10 mm) is desirable despite its higher susceptibility to blackening and contamination. When virus or bacteria elimination is needed, meristem-tip culture is the preferred option. The explant is then further reduced in size (0.5-1 mm length), leaving a meristematic dome with one or two leaf initials. Meristem cultures have the disadvantage that they may have a higher mortality rate and an initial slower growth.

The explant is placed directly on a multiplication-inducing culture medium. For banana micropropagation, MS-based media are widely adopted. Generally, they are supplemented with sucrose as a carbon source at a concentration of 30-40 g/l. Banana tissue cultures often suffer from excessive blackening caused by oxidation of polyphenolic compounds released from wounded tissues. These undesirable exudates form a barrier round the tissue, preventing nutrient uptake and hindering growth. Therefore, during the first 4-6 weeks, fresh shoot-tips are transferred to new medium every 1-2 weeks. Alternatively, freshly initiated cultures can be kept in complete darkness for one week. Antioxidants, such as ascorbic acid or citric acid in concentrations ranging from 10-150 mg/l, are added to the growth medium to reduce blackening, or the explants are dipped in antioxidant solution (cysteine 50 mg/l) prior to their transfer to culture medium.

Usually two types of growth regulators, a cytokinin and an auxin, are added to the banana growth medium. Their concentration and ratio determines the growth and morphogenesis of the banana tissue. We routinely add 2.25 mg/l 6-benzyladenine (BA) and 0.175 mg/l indole-3-acetic acid (IAA) to the initiation medium.

In most banana micropropagation systems, semi-solid media are used. As a gelling agent agar (5-8 g/l) is frequently added to the culture medium but our preference is for Gelrite (2-4 g/l) because of its higher transparency, allowing much earlier detection of microbial contamination. Liquid media are superior for shoot multiplication, but for maximum plant production and survival *ex vitro,* one culture cycle on semi-solid medium is also needed.

Banana shoot-tip cultures are incubated at an optimal growth temperature of 28 ± 2°C in a light cycle of 12-16 h with a photosynthetic photon flux (PPF) of about 60 $\mu E/m^2 s^1$.

Stage 2: Multiplication of shoot-tip cultures

The formation of multiple shoots and buds is promoted by supplementing the medium with relatively high concentrations of cytokinins. In banana, BA is the preferred cytokinin and is usually added in a concentration of 0.1-20 mg/l. For the multiplication of propagules, we use the same medium as for the initiation of shoot cultures (p5 medium containing 2.25 mg/l BA and 0.175 mg/l IAA). If the production of highly proliferating meristem cultures is required, a tenfold higher concentration of BA is added to the culture medium (p4 medium containing 22.5 mg/l BA and 0.175 mg/l IAA). Higher concentrations of the cytokinin BA tend to have an adverse effect on the multiplication rate and morphology of the culture and should therefore be avoided. The rate of multiplication depends both on the cytokinin concentration and the genotype. In general, shoot tips of cultivars having only A genomes produce 2-4 new shoots, whereas cultivars having one or two B genomes produce a cluster of many shoots and buds at each subculture cycle. Approximately 6-12 weeks after culture initiation, depending on the initial explant size, new axillary and adventitious shoots may arise directly from the shoot-tip explant. Clusters can be separated, trimmed and repeatedly subcultured at 4-6 week intervals.

Stage 3: Regeneration of plants

Individual shoot or shoot clumps are transferred to a nutrient medium which does not promote further shoot proliferation but stimulates root formation. The cytokinin in the regeneration medium is greatly reduced or even completely omitted. Within 2 weeks, shoot tips develop into unrooted shoots. To initiate rhizogenesis IAA, NAA (a-naphthalene acetic acid) or IBA (indole-3-butyric acid) are commonly included in the medium at between 0.1 and 2 mg/l. We use the same auxin concentration as in the proliferation medium (0.175 mg/l IAA), but a tenfold lower BA concentration (0.225 mg/l). For some genotypes (*Musa* spp. ABB and BB group) that produce compact proliferating masses of buds, activated charcoal (0.1-0.25%) is added to the regeneration/rooting medium to enhance shoot elongation and rooting. After rooting, plants are hardened *in vitro* for 2-4 extra weeks on the regeneration/rooting medium prior to transplantation to soil.

Conservation of Shoot Cultures

Shoot or meristem tip cultures are suitable not only for the large-scale production of uniform and vigorously growing propagules for field establishment, but also for germplasm conservation. For banana species which are predominately propagated vegetatively, *in vitro* techniques are thus complementary to field conservation.

Banana species can be stored under normal growth conditions at 28°C, but this involves transfer every 2-4 months. Hence there is the risk of losing material through microbial contamination or through human error (mislabelling of culture vessels). Nevertheless, conservation of germplasm under short-term storage conditions is most useful, for instance, to breeders or researchers who wish to maintain a small working collection of breeding material during evaluation, testing, selection and hybridisation, or to maintain products of *in vitro* manipulation. Slow growth conservation has the benefit of reducing the number of subcultures, thus making significant savings in labour input. Germplasm still remains readily available for regeneration, multiplication and distribution. Minimal growth is clearly useful for conservation of genotypes, but is constrained by the risk of genetic changes (somaclonal variation) resulting in the loss of distinct genotypes.

Suppression of growth can be achieved by various modifications of the physical and/or chemical tissue-culture environment. The most common and widely applied growth-retarding factor in banana is low temperature. This technique has been used routinely for many years for the conservation of the International *Musa* Germplasm collection at the INIBAP (International Network for the Improvement of Banana and Plantain) Transit Centre at K.U.Leuven. Banana shoot cultures are stored at 16 ± 1°C. Each accession in the collection is represented by a set of 20 replicates of shoot cultures grown on MS medium, supplemented with 30 g/l sucrose, 2.25 mg/l BA and 0.175 mg/l IAA. Subculture intervals are extended to about 12 months, although large differences in storage potential have been observed among the different genotypes: some genotypes keep for about 20 months whereas others require subculturing every 2 months.

Alternatively, the use of osmotica in banana tissue cultures has been investigated. The addition of 4% mannitol and 3-6% sucrose to the growth medium resulted in 50% growth reduction. Cultures grown on mannitol-enriched medium at 27°C could be kept for 6 months, whereas cultures on the control medium (3% sucrose) started to deteriorate after 4 months of incubation.

Contamination

Contamination in tissue cultures may be caused by endogenous bacteria that escape initial disinfection or by micro-organisms introduced during tissue-culture manipulations. Both types of contaminants may survive in the plant material for several subculture cycles and over extended periods of time without expressing symptoms in the tissue or visible signs in the medium.

'Internal' bacteria are a considerable source of concern in all aspects of plant cell, tissue and organ culture because they hinder the international exchange of germplasm or become a nuisance when contaminated tissues are used as explant material for cryopreservation or for the initiation of embryogenic cell suspensions. It is therefore important that control measures are taken at every tissue-culture step. At the INIBAP Transit Centre, plant material is tested for endophytic bacteria on a broad spectrum bacteriological medium at tissue culture initiation and during annual subculturing. Prior to placing each shoot-tip on culture medium, the base of the explant is streaked onto Difco Bacto nutrient agar, enriched with glucose (1%) and yeast extract (0.5%) in Petri dishes. Testing reveals the presence of cryptic contaminants in

5% of the stored germplasm. In mass propagation systems, positive stock materials should immediately be destroyed. However, germplasm in collections or cultures that are not readily replaceable with fresh material (*e.g.* from field or from greenhouse stock plants) are 'cleaned up'.

Some antibiotics are successful in controlling bacterial contaminants in banana tissue cultures. Rifampicin (100 mg/l) added to liquid cultures for 10-30 days was found most suitable for controlling frequently occurring Gram-positive bacteria in banana shoot tips without affecting plant growth.

However, culturing small meristem tips (1 mm) isolated from contaminated *in vitro* plants or from greenhouse plants obtained from contaminated *in vitro* plants was found to eliminate any bacterial contaminant.

Somaclonal Variation

Shoot-tip culture preserves genetic stability much better than callus or cell suspension cultures, yet somaclonal variation appears to be widespread among plants regenerated from banana shoot-tip cultures.

Off-type frequencies vary from 1% to 74%. The phenomenon of somaclonal variation in the plantain subgroup (*Musa* spp. AAB group) was extensively studied.

It revealed that the incidence of somaclonal variation is strongly influenced by the genetic stability of each cultivar, and that its frequency is amplified by culture-induced factors.

There is no evidence that growth regulators routinely used in tissue culture directly affect the rate of variation, but it has been found that the rate of somaclonal variation is positively related to the generation number.

For 'Williams' (AAA) *in vitro* plants obtained after one and five subculture cycles, dwarfism and leaf-off types counted for 3.7% and 0.7% respectively after one *in vitro* cycle and increased to 6.1% and 1.9% respectively after five *in vitro* cycles.

It is therefore recommended that the number of subculture cycles should be limited to ten or that the number of plants produced from a primary explant should be limited to no more than 1000.

3

Plant Cells

THE STRUCTURE OF PLANT CELLS

The typical cell of the higher plantsis a tiny compartment enclosed by a tough elastic wall.

The wall of cells consists of two major parts:

1. The middle lamella and
2. The primary wall.

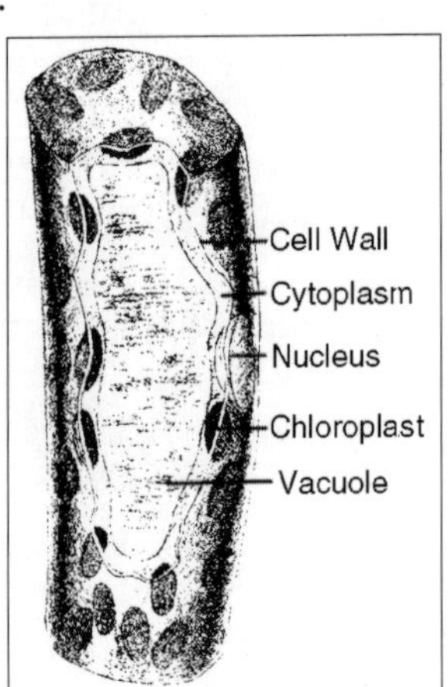

Fig. Perspective View of a Palisade Cell from the Leaf Mesophyll.

In walls of many plant cells a third structural component, the secondary wall, is also present.

Although some plant cells are known which do not have a well defined cell wall, this structure is so generally present in plant cells as to be considered one of their characteristic features. Lining the interior of the wall and occupying more or less of the cell cavity-nucleus is the protoplasm. The protoplasm of active cells is a transparent, slightly viscous, granular material that any conspicuous structural background. It is not homoge-neous, however, and vacucle contains a number of definite structures. One of these, the nucleus, is a denser body which is more or less pheroidal in shape and is separated from the remaining protoplasm by a definite membrane, the nuclear membrane.

Perspective view of a rounding the nucleus are:

- A clear palisade cell from the leaf mesophyll. liquid known as the nuclear sap,
- A delicate network of denser material, the reticulum, and
- One or more small spherical masses of material known as the nucleolus or nucleoli.

All of the protoplasm outside of the nucleus of the cell constitutes the cytoplasm. In a typical mature plant cell the cytoplasm is present as a thin layer lining the inner surface of the cell wall. The two boundary layers ofthe cytoplasm that in contact with the cell wall and that in contact with vacuole are called the cytoplasmic membranes. Imbedded in the cytoplasm are numerous well differentiated bodies known as plastids.

Plastids are specialized cytoplasmic structures which are usually centres of certain types of physiological activity. They are commonly classified on the basis of their colour into three groups: The leucoplasts which are colourless, the cizioroplasts which contain the green chlorophyll pigments (also yellow pigments), and the chromoplasts which contain red or yellow pigments. Chondriosomes, minute rod-like or granular bodies, are also found in the cytoplasm.

The significance of these structures is not positively known, although a number of different roles have been ascribed to them.

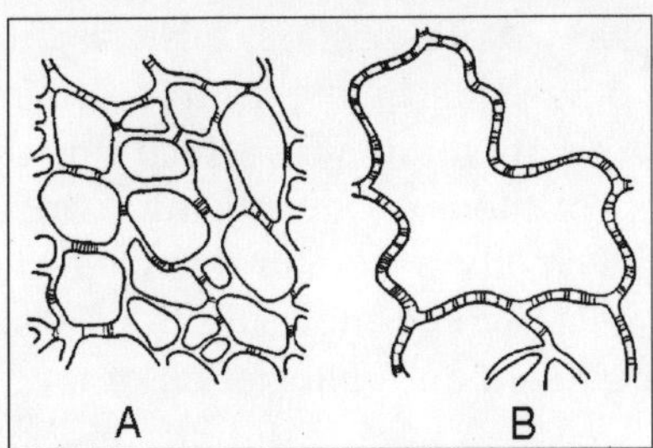

Fig. Plasmodesms in Cell Walls of Tobacco: (A) Sieve Tubes and Companion (B) Epidermal Cells of Leaf.

Although the cell wall appears to imprison each protoplast, and to effectively isolate it from the protoplasm of adjoining cells, actually there is probably a continuation of protoplasm from cell to cell. By certain techniques it can be demonstrated that minute pores extend from cell to cell through the cell walls. These pores often contain cytoplasmic strands which connect the cytoplasms of adjacent cells. These strands are termed plasmodesms.

Livingston has demonstrated the occurrence of plasmodesms in the walls of cells from a number of different tissues of the tobacco plant, and it is generally supposed that they are of widespread if not universal occurrence inplant cell walls. The bulk of the interior of mature plant cells is occupied by a single large cavity, the vacuole, which is filled with cell sap. The cell sap is composed of water in which a great variety of substances are dissolved or colloidally dispersed. There is no general agreement among cytologists regarding the exact classification of the parts of plant cells. The vacuole, for example, is frequently classified as a part of the protoplasm because it first appears as minute droplets in the protoplasm of very young cells. Physiologically, however, the vacuole of the mature cell is as distinct an entity as the protoplasm or cell wall and it is therefore considered as a separate part of the cell in this book.

The following classification includes the principal parts of a mature plant cell. A few of these parts, as for example the various kinds of plastids, do not occur in every cell. Middle lamella, Cell Wall, Primary Wall, Secondary Wall, The Mature Plant Cell, Protoplasm, Cytoplasm, Nucleus, Vacuole, Plasmodesms, Cytoplasmic membranes, Undifferentiated cytoplasm, leuco plasts, Plastids, chloroplasts, chromoplasts, Chondriosomes nuclear membrane, Nuclear sap Reticulum Nucleolus Water, Various compounds in solution or in a state of colloidal dispersion.

ORIGIN AND DEVELOPMENT

As soon as it became clear that all plant and animal tissues were composed of cells the question of the origin of cells naturally arose. This proved to be a difficult problem for the pioneer investigators and for many years there was much disagreement over the question. The now universally accepted principle that cells can arise only by the division of pre-existing cells was first demonstrated beyond any reasonable doubt by Niigeli just before the middle of the Nineteenth Century. A number of methods are now known by which this division is accomplished but a detailed discussion of them is beyond the scope of this book. In higher plants cell division occurs chiefly in certain restricted regions called meristems. The production of new cells involves not only the division of pre-existing cells, but the subsequent enlargement and maturation of their cell progeny.

FORMS AND SIZES

All newly formed cells do not differentiate morpholo-gically in the same way. Some elongate parallel to the axis of growth more than in other directions and thus produce the longer fibre cells that make up much of the xylem and phloem tissues. These cells commonly develop walls that are greatly thickened. Some cells enlarge about equally in all directions forming isodiarnetric cells, the walls of which are never greatly thickened. Cells of this type are present in the pith, leaf mesophyll and in other parenchymatous tissues.

According to Lewis both plant and animal cells are basically tetrakaidecahedrons; *i.e.*, 14 sided although many other geometrical shapes are found, especially in specialized types of cells. An almost endless variation in cell shapes and sizes may be seen in the tissues of any vascular plant, all of which may develop from similarly shaped meristematic cells. In size cells show an equally great range. Most plant cells have diameters that fall somewhere between 10 and 100. A single cubic centimeter oftissue may, therefore, contain millions of cells. Some cells, however, are much larger. Such cells have lengths that are many thousand times their diameters.

CELL WALL

One of the most important features of plant cells is the presence of a conspicuous cell wall that encloses the living protoplasm. There are great variations in the thickness of the walls of different kinds of cells and even greater differences in the physical and chemical properties of cell walls.

In fact the walls of plant cells exhibit such great contrasts in structure and chemical composition that it is difficult to single out any specific cell wall as having a structure that may be considered typical of plant cell walls in general. There are, however, certain features of the cell walls of the vascular plants that are almost invariably present. Probably the most important of these is the almost universal presence of cellulose as the structural framework of the wall.

A second feature of plant cell walls is that the cellulose seems invariably to possess a crystal-like structure. No plant cell wall however, is composed solely of cellulose. There is present in every wall greater or lesser quantities of one or more other substances in addition to cellulose. The most important of these are the pectic compounds, lignin,hemicelluloses, cutin, and suberin.

A third characteristic of most cell walls is their lamellate structure. Most cell walls seem to be an aggregation of numerous delicate lamellae of varying physical and chemical properties, all of which are firmly welded together forming the superficially homogeneous wall.

STRUCTURE OF CELL WALL

The basic unit in the structural organization of the cellulosic framework of the plant cell wall is the cellulose molecule. Cellulose molecules are not units of definite molecular weight but long chains of varying length formed by the condensation of at least a 100-120 and possibly many times this number of /3 d-glucose molecules1. Studies of cellulose walls with the X-ray and polarized light have furnished evidence that the molecules of cellulose are aggregated into bundles known as micelles.

The micelles have been estimated upon the basis of X-ray photographs to have a diameter of approximately 5 or 6 mp and a length of about 6o m. A unit of this size would consist of about sixty parallel cellulose chains each being made up of about 120 glucose units. It is probable that in many cell walls long chain-like molecules formed by the condensation of other sugars are associated with the cellulose chains in the micelles. Originally the micelles were believed to be well defined units which were cemented together by some non-crystalline material. Recent work indicates however, that many of the cellulose chains are much longer than the micellar aggregates and extend from one micelle to another, thus welding the micelles into a coherent an astomosing system.

It is also probable that the micellar aggregates vary in size. Cellulose walls are no longer regarded, therefore as a system of discrete crystalline units bound together by some cementing substance but rather as a structure composed of molecular aggregates are welded together by interlocking cellulose molecules.

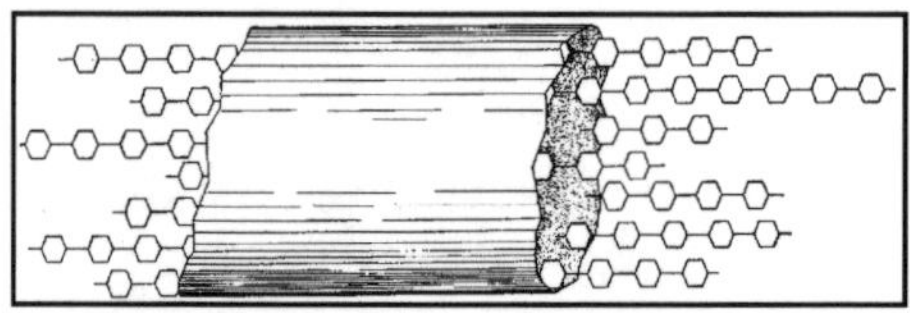

Fig. Diagram Illustrating Structure of A Small Portion of a Cellulose Micelle.

There seems to be little doubt regarding the presence of intermicellar spaces. The spaces form an interconnecting system between the anastomosing micelles.

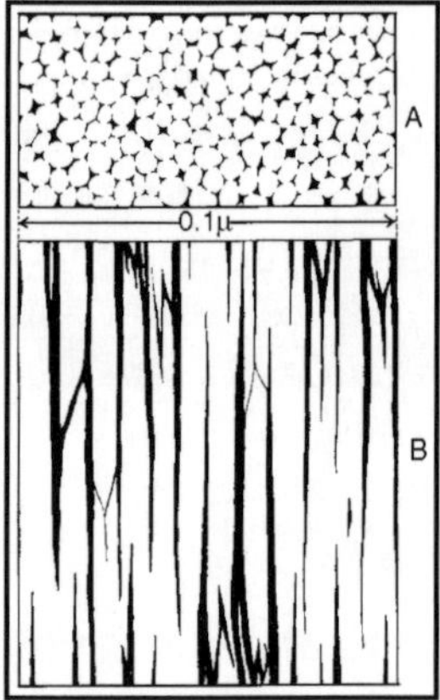

Fig. Micellar Structure of Walls.

Black areas represent intermicellar spaces; white areas represent the cellulose micelles.

(A) Cross section,

(B) Longitudinal section.

In walls that are almost pure cellulose, such as the secondary wall of cotton fibres, it is possible that the intermic-ellar spaces are filled chiefly with water.The smallest visible units of cellulose walls are delicate thread like strands or fibrils. In primary walls these fibrils form a loose anastomosing network the meshes of which are usually filled with colloidal pectic compounds.

In secondary walls the fibrils are often grouped into coarser strands which wind around the cell in a steep spiral the angle of which may vary indifferent layers and even in different parts of the same layer.

PHYSICAL PROPERTIES OF CELL WALLS

The physical properties of primary cell walls differ from those of secondary walls. These differences may be ascribed to the greater abundance of cellulose in secondary walls and to differences in the structural organization of the cellulose in the two kinds of walls. Both primary and secondary walls are transparent to wavelengths of the visible spectrum and both are usually quite pcrmeable to most substances dissolved in water that compare favourably with those of steel.

The breaking strength of a flax fibre, for example, may be as great as 110 kg. per mm.2 of wall area while the tensile strength of a hardened spring steel ranges between 150 and I 70 kg. per mm2. Although the tensile strength of primary walls is considerably less than that of secondary walls it is only rarely that they are subjected to strains in excess of their breaking strength. Primary cell walls are usually quite elastic.

The mesophyll cells of the leaves of some species, for example, are known to undergo reversible changes in volume of 30 per cent or more in response A to changes in turgor pressure. Secondary walls, on the other hand, are elastic only within very narrow limits and break suddenly when their tensile strength is exceeded. The elasticity of primary walls is undoubtedly related to the loose open mesh work of cellulose strands that compose it, while the close parallel orientation of the cellulose micelles in secondary walls prevents any appreciable elongation without rupture.

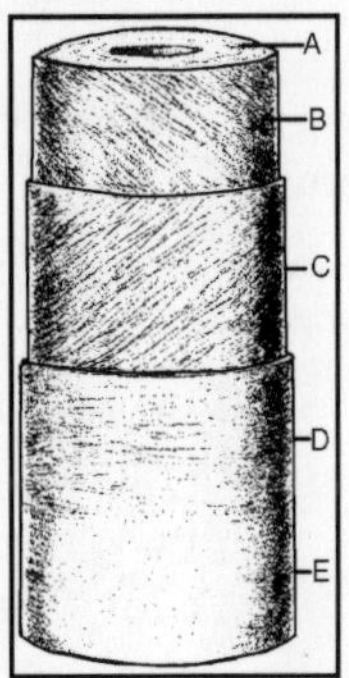

Fig. Diagram Illustrating the Structure of a Thickened Plant Cell Wall.

ORIGIN AND DEVELOPMENT OF THE CELL WALL

In the vascular plants the division of a plant cell is preceded by the division of the nucleus. Nuclear division is a complicated process involving organization of the nuclear reticulum into unit chromosomes, splitting of those chromosomes, migration of one of the chromosomes of each pair to opposite ends of the cell, and finally, reconstitution of a daughter nucleus from each set of daughter chromosomes.

Just after the final stage (telophase) of mitosis a membrane develops in the centre of the cell and extends its margins until it makes contact on all sides with the existing cell wall, thus forming a septum which separates the protoplasts of the newly formed cells. This first membrane, the middle lamella, does not contain cellulose in quantities that permit its detection but seems largely, if not entirely, composed of colloidal pectic compounds. During the enlargement of the cell a thin layer composed largely of cellulose is deposited from the cytoplasm of each young cell on the two sides of the middle lamella.

This primary wall contains large amounts of pectic materials in addition to cellulose and may vary in thickness and in physical properties at different times of the year. Primary walls may become considerably thickened as is the case in the collenchyma cells of young stems. As long as a cell wall possesses a marked elasticity and exhibits reversible changes in thickness or in other physical properties it is classed as a primary wall.

Fig. Diagram Illustrating Formation of a Plant Cell Wall.

(A) Cell plate between two daughter nuclei,

(B) Cell plate(Middle Lam-ella) completely separating the daughter cells,

(C) Primary wall has been formed on each side of the middle lamella. One of the newly formed cells has undergone considerable enalrgement,

The walls of cambium cells, parenchyma cells and collenchyma cells are all examples of primary cell walls. After the enlargement of the cell ceases the deposition of cellulose on the inner surface of the primary wall may continue until the wall becomes conspicuously thickened. When, as in most cases, this thickening is accompanied by an almost complete loss of the elasticity of the wall the added material is known as the secondary wall. Secondary walls are further characterized bythe absence of reversible changes in thickness and by their low content of pectic compounds.

They are largely composed of cellulose which is frequently associated with lignin, hemicelluloses, or other membrane substances. In extreme cases secondary thickening may continue until the wall occupies most of the interior of the cell. The formation of a secondary wall prevents any further enlargement of the cell. All cells have a middle lamella and a primary wall, but secondary walls are present only in certain types of cells. One of the newly formed cells has undergone considerable enlargement. Phloem fibres, stone cells, tracheids, and wood fibres are typical examples of cells with prominent secondary walls.

Increase in thickness of the wall usually appears to take place by the addition of definite layers of cellulose or other cell wall constituents to the inner surface of the existing wall. In most walls these layers are too thin to be detected without swelling or otherwise treating the wall.

CHEMICAL CONSTITUENTS OF CELL WALLS

Cellulose is by far the most abundant compound found in the cell walls of the higher plants. Associated with cellulose in all the primary cell walls of vascular plants are greater or lesser amounts of pectic compounds. Their presence in the middle lamella and in the intermicellar spaces of the primary walls has already been noted. Lignin is an important constituent of the walls of most of the cells that make up woody tissues and it also occurs commonly in other thickened walls.

Although lignin can be isolated from cell walls by suitable treatments as a brownish amorphous substance it is probably altered considerably in the process. Its structural formula and molecular weight are unknown. Since the lignins produced in different species appear to differ in their chemical properties it seems very probable that lignin is not a compound of definite molecular weight but a complex mixture of a number of chemically similar substances. The lignin of spruce wood has been extensively studied and several workers have suggested that its molecular weight lies between 800 and 900.

Lignin first appears in the middle lamella and primary wall and later canbe detected in the secondary wall. When associated with cellulose, lignin is present in the spaces between the micelles and not within the micelles. Lignifled walls are usually freely permeable to water and solutes. The tensile strength of lignified cell walls is the same as that of cellulose walls but lignified walls resist compression better than cellulose walls. The increased resistance of lignified walls to compression is explained by the assumption that the presence of lignin in the intermicellar spaces welds the cellulose micelles into a single coherent mass and thus prevents the bending and buckling of the cellulose strands when they are subjected to compression strains.

Cutin is the name applied to the mixture of wax-like materials found on the outer surface of the epidermal cell walls of leaves, stems, fruits, and other organs. These wax-like substances are intimately associated with cellulose and often with pectic substances producing a wall of great structural complexity. Cutinized cell walls are relatively impermeable to water. The presence of cutin in the outer walls of epidermal cells greatly reduces the evapouration of water from the surfaces of plant tissues. Subcrin is similar in many of its properties to cutin.

It constitutes an important part of cork cell walls and it is also found in the walls of a few other specialized types of cells. Most of the surface of perennial plants, aside from the leaves and very young stems is covered with suberized cell walls. Such walls are relatively impermeable to water. Hemicelluloses are a poorly defined group of poly saccharides associated with cellulose in plant cell walls. They are not chemically related to cellulose, as the name implies, but possess very different chemical and physical properties. Callose is the name given to a carbohydrate membrane substance found in the perforated septa (sieve plates) of the sieve tubes. Similar material presumably of the same chemical composition, has been found in pollen grains and constitutes the inner layer of pollen tubes.

It has also been reported as occurring in the fungi. The exact chemical composition of callose is unknown since it has never been obtained in sufficient amounts to permit quantitative determinations. Chitin, a nitrogenous substance common in the exoskeleton of insects, is a constituent of the walls of many fungi and bacteria. It has been reported as being present in the walls of certain algae but it is unknown in any of the higher plants. Tannins are commonly found in the cell sap but they also occur in the walls of certain tissues, especially cork and wood cells. Mucilages are common constituents of the outer walls of many water plants and occur also in the outer walls of some seed coats, in glandular hairs and in other specialized tissues. Inorganic compounds such as silica and salts of calcium, iron, and other metals are also present in some plant cell walls. None of these inorganic compounds, however, are regarded as essential constituents of the cell wall.

PROTOPLASM

In most mature cells of the vascular plants protoplasmis present only as a thin layer covering the inner surface of the cell walls but in some specialized cells branching strands of protoplasm also extend across the vacuole. Under high magnification the protoplasm of active cells appears as a colourless fluid, in which are suspended numerous tiny granules and droplets of insoluble materials. These granules frequently exhibit active Brownian movement. The fluid component of protoplasm also is frequently in motion, streaming around the inner surfaces of the cell walls. Embedded in this fluid component, and carried passively by it, are the specialized bodies known as plastids.

Although protoplasm appears to be a simple liquid, no simple liquid could possibly possess the remarkable powers of synthesis, assimilation, reproduction growth and sensitivity that characterize the protoplasm of living plant cells.The properties and behaviour of protoplasm clearly show that it is not a substance but that it must be regarded as a complex system of substances.

This system is dynamic; it is constantly undergoing changes yet at the same time the changes are so regulated and controlled that the system is not disrupted. A cell is alive only so long as the organization of this dynamic protoplasmic system is maintained.

The protoplasmic system of living plant cells almost invariably contains a well differentiated globular body known as the nucleus. All of the protoplasm outside of the nucleus is designated as cytoplasm. Although the nucleus is separated structurally from the cytoplasm by a delicate membrane the two components are not physiologically isolated. The nucleus and the cytoplasm appear to be mutually dependent and there is good reason to believe that the nucleus controls and regulates the physiological processes that occur in the cytoplasm. Probably because of the close relationship between the nucleus and the cytoplasm many biologists use the terms protoplasm and cytoplasm synonymously.

THE CHEMICAL COMPOSITION OF PROTOPLASM

Since protoplasm is a dynamic system of substances it is not possible to subject it to chemical analysis without destroying it. In the strict sense, therefore, it is impossible to discover the chemical composition of protoplasm. It is possible, however, to examine the substances present after the protoplasmic system has been destroyed and to determine their chemical composition and relative abundance. A number of such studies have been made.

Water is the chief component of all physiologically active plant protoplasm usually making up more than 90 per cent of the system. The water content of the protoplasm of dry seeds, on the other hand, may be less than 10 per cent. Most attempts to determine the chemical composition of plant protoplasm have been made upon species of the myxomycetes. At certain stages in their life history these organisms consist of naked masses of labile protoplasm. They are often found flowing over rotten logs in damp woods.

The fact that the myxomycetes provide relatively large quantities of protoplasm entirely free from cell wall material has made them a favourite object for chemical analysis. Even in such organisms, however, not all of the constituents of the plant body can be regarded as integral parts of the protoplasm. Distributed throughout the protopl-asmic mass are particles of foods and other inert materials which cannot be separated from the protoplasm. As shown in this analysis proteins and other nitrogen-containing compounds constitute the bulk of the organic matter in the plasmodium of this species. Many different varieties of proteins are known to occur in the protoplasm of plant cells. They are compounds of enormous molecular weight and undoubtedly make up a large proportion of the labile structural framework of the protoplasm. Lipids constitute a smaller fraction of the protoplasm than the proteins. Three types of lipids occur in the protoplasm; the true fats (oils), the phosphatides (phospholipids) and the sterols, of which phytosterol is an example. The oils are generally suspended in the protoplasm in the form of minute globules. They are probably more important as food reserves than as actual constituents of the protoplasm. The phospho-lipids and sterols, on the other hand, are believed to be essential constituents of the protoplasmic system.

The water-soluble carbohydrates, amino acids, etc., present in the plasmodium of this species are probably almost entirely foods. The inorganic compounds (mineral matter) in plant cells are chiefly the phosphates, chlorides sulfates, and carbonates of magnesium, potassium, sodium, and calcium. It supplies no more information regarding the organiz-ation of protoplasm, however, than a chemical analysis of the ground up debris of a wrecked house would furnish regarding the structure of that house. This point is worthy of emphasis, since with the progress of biology it becomes clearer and clearer that the properties of protoplasm are as much a function of its physiochemical organization as of the specific kinds of compounds present.

Although we commonly refer to protoplasm as the essential constituent of all living cells, it is evident that there are at least as many different varieties of protoplasm as there are species of plants and animals. All of them, however, are dynamic systems composed of the same types of compounds, and all of them possess a colloidal organization of a complex type.

THE MECHANISM OF PLANT CELL GROWTH

The plant cell after formation normally increases very considerably in size. This process may be studied in two situations:

- When the cells are in intact organs,
- When they are in isolated systems.

The first indicates the scope and characteristics of the process, the second the factors that control the several aspects of it. The two groups of studies based on the different situations complement each other, and it is therefore proposed to consider below, first, observations made with intact organs, and secondly, observations made with isolated fragments. This treatment of plant cell growth is not intended to be exhaustive and one topic in particular is omitted.

For two reasons no discussion is attempted of the nature of the auxin effect in cell growth. The subject is omitted, since it has recently been discussed in some detail by Audus and, secondly, since it is possible that an elucidation of this problem can only be developed after the nature of cell growth has been more fully explored.

Hitherto plant cell growth has frequently been described as cell extension. In our view this term is an unfortunate one. It is probably a legacy of the period when cell growth was considered a simple matter of an inflation due to the absorption of water. Recent work, however, has shown that the increase in size of the cell involves increases in many components of the system, and we therefore prefer the term cell growth to the older one of cell extension.

CELL GROWTH IN INTACT ORGANS

The process of cell growth involves the changes that occur between the meristematic and the mature fully vacuolated state. The meristematic cell is small isodiametric and without a prominent central vacuole. As the cell enlarges a central vacuole develops and the shape changes.

Eventually a large parenchymatous cell is established with a thick wall and a thin peripheral layer of cytoplasm. This set of changes is readily studied on the root. It is probable that this is a more satisfactory experimental object than the coleoptile, since with this organ the earlier stages of growth are not readily accessible to investigation.

The root, being an organ of unlimited growth, has an apical meristem from which cells are being continuously generated. These cells immediately they are formed commence enlarging, and thus at increasing distances from the apex the tissue is composed of cells in progressively advanced stages of development. A similar situation is involved in the apex of the stem. Here, however, the position is complicated by the development of lateral members from the surface of the meristematic dome.

In the root the surfaces of the growing regions are not covered with developing primordia, and this circumstance has been exploited in the development of two sets of techniques for the study of cell growth. Since the surface of the growing region is naked in the root direct observations may be made microscopically on epidermal cells.

The dimensions of epidermal cells may be measured at increasing distances from the apex and the growth of these cells determined over the whole range of development. This technique has been used in principle by Burstrom, Brumfield and by Goodwin.

It has been shown that sliding growth is not involved, and changes in surface cells may therefore be taken as representative of corresponding changes occurring in deeper tissues. This general technique has yielded important results, and it has the merit that it involves observations on a more or less uniform tissue.

It carries the limitation, however, that only a limited variety of observations can be made with it. It cannot, for instance, yield data regarding metabolic conditions in developing cells. In order to avoid this difficulty another experimental approach has been proposed by Brown and Broadbent.

These workers cut successive sections from the apex backwards, made observations on each section, and related the values obtained to the total number of cells in each section. They used a technique for determining the number of cells in a tissue originally developed by Brown and Rickless.

The range of observations that can be made with this technique is clearly greater than that possible with the earlier procedure. It involves, however, two possible sources of error. Cell division does not cease at the same distance from the apex in all regions of the root, and all cells in the section are therefore not exactly at the same stage of development. Secondly, sectioning necessarily involves a destruction of cells the extent of which cannot easily be estimated.

The first source of error is not decisive, since the distance over which cell division ceases is small relative to the total distance over which growth occurs, and the average given by the section therefore represents approximately the position for all the cells in the section.

The error from the destruction of cells can be reduced by increasing the length of the section and thus decreasing the proportion of damaged cells. These data are from the observations of Brown and Broadbent. The full development occurs over a distance of about 5.0 mm. from the apex and involves about a twenty- to thirtyfold increase in volume. The enlargement is accompanied by important metabolic changes. In the present connexion it may be noted that the enlargement in the early stages is accompanied by the gradual suppression of the capacity to divide.

The zone that stretches from the tip to about 1.5 mm. behind it is that over which mitotic figures are normally distributed. At the same time the observations of Wagner and of Gray and Scholes have shown that the mitotic frequency is greatest at the tip and decreases progressively with increasing distance from it.

Brown has emphasized that the frequency of mitotic figures may not be an index of the relative rate of division. In this case, however, since the frequency decreases to zero it probably is, and the data therefore suggest that after formation at the apex the increase in the average cell size is accompanied by a corresponding increase in the average length of the interphase. As growth proceeds the probability of division decreases, and at a certain point the probability becomes so small as to constitute virtual cessation of the process.

The conditions that determine this transition have not been explored, but it is clear that the fact that it occurs indicates that growth in the early stages is accompanied by metabolic changes that tend to suppress division. At about the point at which division ceases a prominent central vacuole appears in the cell. This vacuole, which is probably a development from a dispersed vacuolar system in the meristematic cell, increases in size with the enlargement of the cell.

During growth, however, the percentage water content is increasing, indicating that the volume of the vacuole is increasing more rapidly than the other components, and that the enlargement of the cell is being determined primarily by the absorption of water. At the same time it is evident that the increase in water content is being sustained by relatively smaller increases in other phases of the system. These data indicate that the dry weight of the cell increases about tenfold during the course of growth. Preston and Clark and Wirth have shown that the cellulose content of cells in the coleoptile increases during growth and Wirth has demonstrated corresponding changes with soluble carbohydrates. The increase in dry weight of the cell in the root is therefore probably due partly to similar changes. It is evident, however, that it is not due only to changes in these constituents. In 1941 Blank and Frey-Wyssling described a series of important observations on cell growth in the coleoptile of maize. They reported an increase in protein content during the enlargement of the cell. Blank and Frey-Wyssling found similar changes occurred during cell enlargement in the hypanthium of *Oenothera*. Kopp calculated that in the root the protein content increased, and Brown and Broadbent demonstrated this condition experimentally in pea roots.

The appropriate curve is reproduced from the data of Brown and Broadbent, and it is evident from this that protein increases progressively during growth of the cell, resulting in at least a fivefold increase.

At the same time it may be noted that when growth ceases there is a slight but significant decrease in protein, the final level being lower than it is when growth stops, although greater than it is in the meristematic region. The increase in protein necessarily implies an increase in the general metabolic activity of the cell during growth. Kopp calculated and Brown and Broadbent later demonstrated an increase in respiration rate per cell during growth.

The data of Brown and Broadbent suggest, however, that when growth ceases there is a slight fall in respiration which is coincident with the decrease in protein. The increase in protein content might also be expected to increase the quantities of particular enzymes. We have recently been examining the changes in the activities of certain enzymes during the growth of the cell.

These data have been obtained by taking successive sections along bean roots, killing the sections by freezing at - 10°C., determining on the killed tissue the activities of certain enzymes, and finally reducing the values for each section to a unit cell basis. The activities of three enzyme systems have been investigated: a dipeptidase which hydrolyses alanylglycine, an invertase, and an acid phosphatase.

The activity of the dipeptidase at increasing distances from the apex of the root of barley has already been examined by Bottelier, Holter and Linderstrom-Lang, and Wanner and Leupold have determined the relative sucrose-splitting capacity in different regions of the root of maize. In the bean roots growing in the conditions of our experiments growth continues over about the first 8 mm. The data show that over this range there is about a fivefold increase in dipeptidase, about a twofold increase in phosphatase, and about a twentyfold increase in invertase activity.

Further, it is evident that in about the region where growth in length is ceasing the activities of the respective enzyme systems decrease. Determinations of protein in cells of bean roots are not available, but in view of the earlier observations with peas it is probable that the increase in enzyme activity is due to the differentiation of part of the protein increment into enzyme bodies, and that the decrease in enzyme is similarly partly due to a decrease in protein. It is an important aspect of the situation that the activities of the three enzymes do not change relatively to the same extent.

The relative increases are different and the ratios of the activities of any two of the enzymes therefore change during the course of growth. A similar position is indicated by some observations of Linderstrom-Lang and Holter on the activities of what may be two distinct enzymes cleaving leucyl-glycine and alanylglycine respectively at increasing distances from the tip of the root of barley.

The data, although they are not expressed in terms of unit cell number, nevertheless show that the ratio between the two enzymes changes as the distance increases. Clearly since the ratios between the enzyme activities change it is evident that the metabolic pattern must change during the course of growth.

During the growth of the cell a differentiation probably occurs in the metabolic state such that the relative intensities of different reactions change in the course of the process. In particular, these observations on the enzyme complement suggest that the metabolic activities of the young cell are different from those of the mature. This conclusion is consistent with the results of an investigation by Morgan of the amino-acid composition of the proteins in different parts of the growing zone of the bean root.

The proteins from sections taken at increasing distances from the apex of the root have been hydrolysed and the hydrolysates examined chromatographically. It is evident that the composition of the proteins varies considerably.

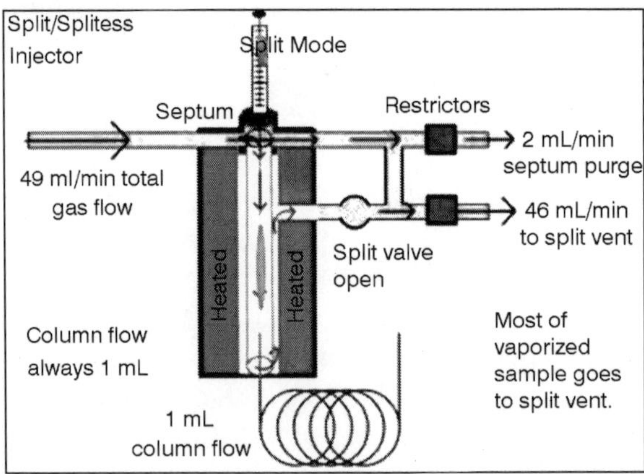

Fig. Chromatographically

Such differences in the composition of the proteins suggest of course corresponding differences in the generalized metabolic pattern. This general conclusion is also consistent with a variety of observations by other workers. Berry and Brock have reported that the respiration of the tip of the onion root is more sensitive to cyanide than is that of the mature tissue, and Kopp has found that per unit protein it is lower at the tip than it is in the mature regions. The data of Brown and Broadbent suggest that the increase in protein per unit increase in volume is greater in the earlier than it is in the later stages of growth.

Finally, Burstrom has shown that the reactions of root cells to sugar, temperature and heteroauxin are different in the early and late stages. The growth of plant cells is interesting on account of the fact that the protoplasm is covered with a solid wall. Any irreversible increase of the cell size involves the growth in area of this wall, which consists partly of crystallized substances. The development of cell walls, especially in the cells of higher plants, passes through two different stages.

As long as there is growth in area, they remain thin and flexible. This fine membrane is called the *primary cell wall*, and it is said to grow in area by *intussusception*. As soon as the final shape and size of the cell are reached, the membrane is strengthened by adding new layers to the primary wall.

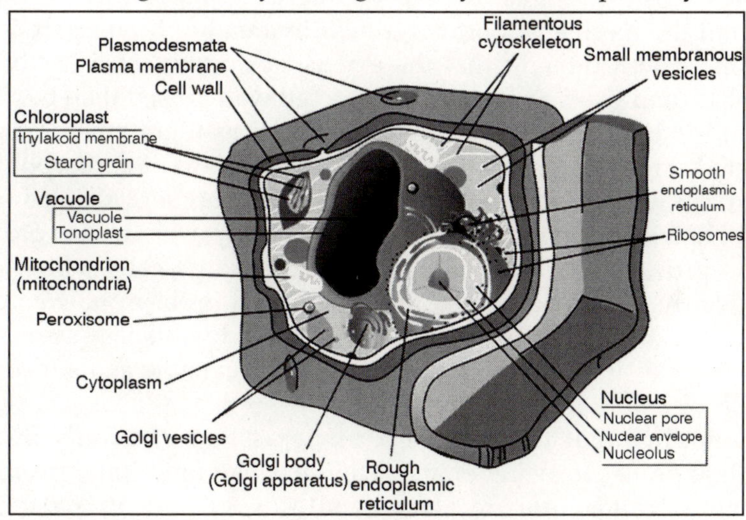

Fig. Plant Cell

During this second stage the cell wall grows in thickness by *apposition*. The combined layers are called the *secondary cell wall*. The thickness of this is often so large that the slender primary wall is overlooked or neglected, especially in materials of technical importance such as wood pulp, textile fibres, etc. Indirect methods of investigation, such as polarized light and X-rays, indicate that the primary wall consists of a loose submicroscopic network of crystallized strands of cellulose or chitin, arranged in a dispersed texture.

In the secondary wall the framework is much more compact. This compactness is obtained by a parallel arrangement of the submicroscopic strands forming a highly anisotropic *parallel texture*. Two different types of growth in the area of the primary wall can be distinguished:

TIP GROWTH

Tip growth of filamentous cells such as fungal hyphae, hairs (*e.g.* root hairs), latex tubes, pollen tubes, etc. In this case, only the tip of the cell is involved in growth, and no intussusception occurs away from the end. Sometimes the tip growth of very long cells (e.g bast fibres of ramie, which reach a length of over 20 cm.) continue after the formation of the secondary wall has already started in the middle body of the cell.

EXTENSION GROWTH OF CELLS

Extension growth of the cells in elongating tissues. In this case the whole length of the primary wall is said to increase in area by intussusception. There is an enormous bibliography on this subject, because the extension growth is stimulated by auxin, which was considered at the time as the hormone of the extension growth.

However, the question how auxins intervene in this process cannot be solved until the mechanism of the growth in area has been cleared up. For a long time it was thought that the extension growth consisted of simple water intake, and the turgor stretch of the cell wall caused thereby.

But it has been shown that this type of cell wall growth is much more complicated, since it involves respiration, biosynthesis, and morphogenesis. Moreover the work needed for the elastic deformation of the growing cell wall is negligible as compared with the total energy available for growth. As extension growth concerns tissues with numerous cells which expand simultaneously at different rates, the geometrical problem how contiguous polyhedra can change their shape without being individualized arises. Krabbe thought that a disintegration of the tissue was inevitable and suggested the hypothesis of *sliding growth*.

This implies that the cells of a tissue assume the individuality of independent unities growing along each other like unicellular protophytes. This is an extreme interpretation of sliding growth it is true, but it demonstrates clearly that such an idea is theoretically unsound.

Priestley published other arguments against the possibility of this type of growth; sliding of the two adjacent cell walls would interrupt the channels of pits, so that their orifices would no longer correspond.

In consequence one might expect to find a large number of unilateral pits in tissues in which sliding growth had occurred. As this is not the case, Priestley postulated that the adjacent cell faces must grow together at the same rate, and he termed this *symplastic growth*. His theory cannot, however, explain how latex tubes and other specialized cells grow across tissues and thereby come in touch with cells which previously were not their neighbours. This frequently occurs when derivatives of the cambial cells, such as tracheids and fibres, elongate in the tissue.

Two adjacent cells are pushed apart by the tip of a cell which intrudes between them. This has been termed intrusive growth by Sinnot and Bloch or interposition growth by Schoch-Bodmer. The last author emphasizes that only the tips of elongating fibres behave in this way. According to her, the walls of the central body of the cell do not grow any longer, so that the entity of the tissue is secured. Pits are only found in that part of the cell wall, and the fibre tips are free from them.

From this description it follows that different parts of the cell wall must behave differently. A plant cell does not grow as a whole, but in a differentiating cell there are growing and full-grown parts at the same time.

STRUCTURE OF GROWING CELL WALLS

The electron microscope shows that young cell walls have a woven texture of dispersed cellulose microfibrils. The diameter of the microfibrils involved measures about 250 A. However, the electron microscope is said to be unsuited for research in cell physiology and growth because the specimens must be examined completely dried *in vacuo*.

Nevertheless, it is possible to take advantage of the results obtained by electron microscopy if they are properly combined with those of other methods. The natural state of the cellulosic framework of the living cell wall can be reconstructed from the electron micrograph if its chemical composition is known. In growing coleoptiles we have found 92.5 per cent water. If this water content is taken as an average for all cell constituents, the cell wall contains only 7.5 per cent dry matter. In corn coleoptiles 32 mm. in length twothirds of this dry material is composed of substances other than cellulose, such as pectins, hemicelluloses, wax. Therefore, the growing cell wall consists of only 2.5 per cent cellulose by weight.

As the density of crystallized cellulose is about 1.55 this is even less by volume. It shows how astonishingly loose is this framework in a growing cell wall, including plenty of space for living cytoplasm and all kinds of incrusting substances.

Only when this highly hydrated, spacious network is freed from non-cellulosic substances and completely dried does it furnish the picture of a dense interwoven texture in the electron microscope. The impression of a rather compact framework is further enhanced by the high focal depth of the electron microscope, which yields a sharp picture of microfibrils situated in rather distant planes.

Whereas it is difficult to understand how a woven texture as shown by the electron micrograph can grow by intussusception, where new microfibrils can readily be interlaced. Another handicap to electron microscopy of cell walls has been the method of preparation, which previously consisted in disintegrating cells in a blendor whereby pieces thin enough for electron transmission were obtained.

From these specimens it was not possible to locate the pieces examined in the original cell wall, and the micrographs obtained did not show whether the whole wall of a cell or only parts of it are involved in growth. Muhlethaler showed, however, that the purified cell walls of macerated meristems are thin enough for direct examination without cutting them into smaller bits. This is possible because in these membranes he cellulose frame represents only about 2.5 per cent by weight of the living wall, which has a diameter of less than 1¼ in the swollen state.

So the dried cell wall freed of all incrusting substances would be only about 0.025¼ thick, if its mass were spread as a homogeneous film. Yet, according to the porous structure of the wall the dried framework is correspondingly thicker, say 0.075 ¼. This would correspond to a woven texture of about three microfibrils thick. As macerated collapsed cells consist of two superposed walls the thickness of the preparation may amount to 0.15¼, which is an appropriate value for the examination in the electron microscope. The possibility of investigating whole cells proved an opportunity for observing the same spot of the wall alternatively in the electron and in the ordinary microscope.

For this purpose the walls are stained with benzoazurin, and the resulting dichroism is examined in polarized light. As the direction of the strong absorption coincides with the main trend of the microfibrils, a mutual check on the observations in the electron microscope and the results of this indirect method is possible.

Thereby an identification of different cell types and the investigation of individual faces of a cell wall are feasible. Our first intention was to elucidate the tip growth in order to observe the formation of new microfibrils at the very tip of the cell where intuesusception growth occurs. *Root hairs* seemed a favourable object for this purpose, because their tip growth is very fast.

This research yielded the discovery that the tip of root hairs (*Zea mays*) is covered with a felt-like layer of cellulosic microfibrils. The felt is more than 1 ¼ thick and corresponds to the slime layer which covers the growth region of the hair. The microfibrils are anchored in the interwoven cell wall. The slimy consistency of this layer is due to a large amount of interfibrillar substances with high swelling power, such as pectins. These can easily be demonstrated by pectin dyes (ruthenium red, methylene blue), whilst microchemical reactions, aiming at proving the presence of cellulose, fail.

This is probably due to the fact that the microfibrils found are so well crystallized that no intermicellar spaces are left for the penetration of cellulose dyes. The tip of the root hair is so woolly on account of this extracellular felt that the formation of new microfibrils cannot be observed. It is probable, however, that the cytoplasm which synthesizes the wall substances soaks this system of cellulose microfibrils, which on account of the hydrated state of the pectins must be very loose. As the process of neoformation of microfibrils is hidden in the root hairs, we looked for objects where the growth zone is not covered by slimes, such as hyphae of fungi and pollen tubes. A favourable object is the growing *sporangiophore of Phycomyces* before the tip is inflated for the formation of the sporangium. The chitinous cell walls of fungi have the same microfibrillar structure as those of higher plants; the only difference is that the microfibrils consist of chitin instead of cellulose. Curiously enough both types of microfibrils have approximately the same diameter and both are arranged in loosely woven dispersed or compact parallel textures.

At the tip of the young sporangiophore the texture of the microfibrils is very dense, and is somewhat obscured by some incrusting substance which is difficult to remove. The same difficulty is encountered with the tip of *pollen tubes*. These are very smooth and do not show any structure visible in the electron microscope. Only after extraction with such a strong leaching agent as 10 per cent KOH can the microfibrils be made visible. Therefore, the new wall formed at the tip seems to be cutinized instantly in order to prevent desiccation of the pollen tube which is growing in air. The microfibrils of pollen tubes have a smaller diameter than those of the microfibrils found hitherto in primary cell walls. The supposition that this might be due to the haploid state of the pollen tubes turned out to be wrong. There is no relationship between the size of microfibrils and the number of genoms in the nucleus.

This was shown by looking at the haploid alga *Spirogyra,* where very coarse microfibrils occur in the cell wall. Incidentally it was found that the rounded wall of the end cell in a *Spirogyra* thread showed an unexpected tip growth. At the very tip, longitudinal microfibrils spread out from the woven texture of the wall into the open, leaving a gap between them. Along the margin of this gap cross-fibrils are woven into the pattern of the cell wall. It seems that the cytoplasm of the cell tip is so to speak naked while forming the new wall.

FACTORS DETERMINING THE OSMOTIC PRESSURES OF PLANT CELLS

Any factor which influences either the water content of a plant cell or the solute content of its cell sap will have an effect on the magnitude of the osmotic pressure of that plant cell. The water content of the plant as a whole, and hence of its constituent cells, is controlled principally by the relative rates of transpiration and the absorption of water.

The latter process is influenced very markedly by the water content and other conditions prevailing in the soil. Individuals of the same species invariably have a higher osmotic pressure when growing under drought conditions than when provided with a favourable water supply. This is at least partially due to the low water content of the leaves which results when the available soil water supply becomes low. Other factors which are also probably involved are a decrease in the growth rate of the plant which often permits an accumulation of mineral salts and soluble foods, and a shift of the starch 1 soluble carbohydrates It should perhaps be emphasized that the term osmotic pressure of a tissue can possess meaning only in the sense of the average osmotic pressure of the cells composing that tissue.

The solute Content of the Cell sap is Controlled by the specific metabolic processes of the plant and by the absorption of mineral salts by the plant from its environment. The rate of photosynthesis is an important factor indetermining the osmotic pressure of plant cells, particularly those of leaf tissues. The influence of the inherent metabolic processes of any species upon the kinds and concentrations of various types of soluble organic compounds produced, such as simple carbohydrates, organic acids, amino acids, etc. has an exceedingly important effect on the magnitude of the osmotic pressure in any species.

Metabolic conditions and their effects upon osmotic pressures may also be altered by environmental conditions. A well-known example of this is the difference in the osmotic pressures of sun and shade leaves on the same plant. The former almost invariably have the higher osmotic pressure due presumably to their greater photosynthetic activity. The mineral salts which contribute to the osmotic pressures of plant cells are all absorbed from the soil solution, or in aquatic species from the water in which part or all of the plant is immersed.

Different species of plants vary greatly in their toleration of high concentrations of mineral salts in the soil. All species can become adjusted within limits to a change in the mineral salt content of the substratum. This adjustment takes the form of an increase in the osmotic pressure of the plant with an increase in the osmotic pressure of the medium from which it obtains its mineral salts.

Other tissues of plants also show an increase in osmotic pressure with increase in the osmotic pressure of the soil solution, but such increases are generally most pronounced in the roots.Species indigenous to alkali soil regions usually have relatively high osmotic pressures.

Alkali soils are rich in soluble salts, and the high osmotic pressures found in species native to such soils are due to the absorption of relatively large quantities of mineral salts. In fact the highest recorded osmotic pressure for any species of plant atmos was found in an alkali soil species, A triplex con fertifolia. Salt marsh plants also generally have relatively high osmotic pressures.

This is also a correlation with the relatively high salt content of the substratum. Diurnal Variations in the Osmotic Pressures of Plant Tissues. The osmotic pressures of plant tissues fluctuate diurnally. Such variations are more marked in the aerial than in underground portions of plants. Their magnitude and direction are very strongly influenced by the environmental conditions to which the plant is subjected.

Daily variations in osmotic pressure have been studied more extensively in the leaves than in any other organ of the plant. The data for the leaves of the prairie grass, Andropogon scoparius were obtained under extreme drought conditions, but similar, although less marked diurnal variations in the osmotic pressure of leaves are undoubtedly of regular occurrence in most species, at least on clear days.

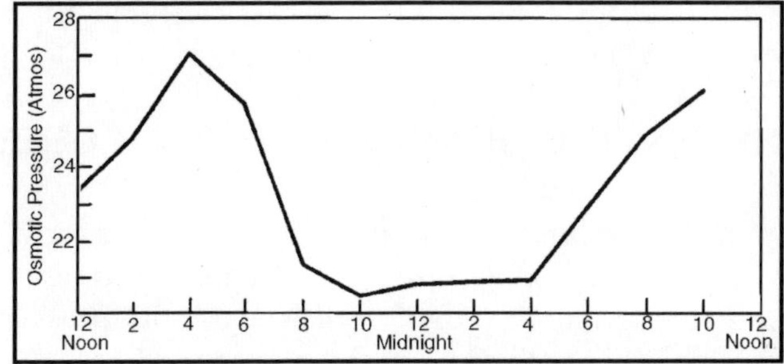

Fig. Daily Variation in the Osmotic Pressure of the Leaves of Andropogon Scoparius

SEASONAL VARIATIONS IN OSMOTIC PRESSURES

The rather consistent increase in the osmotic pressure of leaves which usually occurs during the daylight hours is probably conditioned primarily by two factors: the accumulation of soluble carbohydrates or other organic compounds resulting directly or indirectly from photosy-nthesis, and the decrease in the water content of the cells due to an excess rate of transpiration over the rate of absorption of water resulting in a shrinkage in the volume of the cells.

The minimum leaf osmotic pressure, which is usually attained between midnight and dawn, probably corresponds to the period when the leaf cells are at their maximum water content, and at their minimum organic solute content due to the continuance of translocation out of the leaves during the night. Seasonal Variations in the Osmotic Pressures of Plant Tissues. In temperate regions the osmotic pressures of at least the exposed organs of plants are usually higher in the winter than in the summer.

Seasonal variations in the osmotic pressures of temperate zone plants have been studied principally in the leaves of evergreen species. Such variations are also known to occur in the exposed woody stems of plants, although information about them is much less complete. The increased winter osmotic pressures are undoubtedly correlated principally with an increase in the soluble carbohydrates in the cell sap due to a shift in the starch soluble carbohydrate equilibrium caused by low temperatures and to a lesser extent to the reduced water content of the leaf tissues which is usually characteristic of the winter condition.

In some species the first of these factors is predominant, in other species the second. Similar seasonal variations in osmotic pressure occur in the leaves of evergreen herbaceous species, as for example winter wheat. In many woody species of semi-desert regions the osmotic pressure of the leaves increases markedly during the hot, dry season (May-July) but shows a rapid decrease with the advent of summer rains. The increasing osmotic pressure of the leaves of such species during the dry months is closely correlated with the increasing unavailability of soil water.

THE OSMOTIC QUANTITIES OF PLANT CELLS

Let us suppose three similar cells, each in a state of incipient plasmolysis, and each with a cell sap osmotic pressure of 10 atmos., to be immersed, the first in pure water, the second in a solution with an osmotic pressure of 6 atmos. and the third in a solution with an osmotic pressure of 10 atmos. The volume of the liquid in which each cell is immersed is supposed to be very large in proportion to the volume of the cell. In order to simplify the discussion certain assumptions will be made regarding these cells and, unless stated to the contrary, regarding all other cells.

These are:

- That the cells are at the same temperature,
- That the cytoplasmic membranes are permeable only to water while the cell walls are freely permeable to both water and solutes and
- That the walls of all the cells are equally elastic. Furthermore, changes in the osmotic pressure of the cell sap of these three cells due to the changes in the volume of the cells will be disregarded. Diffusion of water into cell A begins immediately upon its immersion.

This inward movement of water is osmosis. The diffusion pressure deficit of the water in the cell sap is JO atmos. (since its osmotic pressure is 10 atmos and it is not subjected to pressure); that of the surrounding water zero, hence the diffusion pressure of the external pure water is 10 atmos. greater than that of the water in the cell sap. The entrance of water into the cell exerts a gradually increasing pressure against the protoplasm which is in turn transmitted to the cell wall. This pressure is called the turgor pressure and is the cause of the gradual distension of the cell.

As soon as any turgor pressure is exerted against the walls of the cell they exert a counter pressure the wall pressure against the protoplasm and cell sap. The wall pressure is always equal to but acts in the opposite direction to the turgor pressure. The maximum turgor pressure which can be developed by this particular cell is 10 atmos., since the osmotic pressure of a solution is a measure of the maximum turgor pressure which it can develop.

At this point the wall pressure of the cell will also be equal to 10 atmos. The wall pressure of a cell will have exactly the same effect upon the diffusion pressure of the water in the cell sap that a pressure of equal magnitude exerted by a piston would have on water subjected to it. In this cell, therefore, the wall pressure will increase the diffusion pressure of the enclosed cell sap by 10 atmos.

Since the initial diffusion pressure deficit of the cell sap was Jo atmos. and the development of a wall pressure of 10 atmos. has raised the diffusion pressure of the water in the cell sap by JO atmos. the diffusion pressure deficit of the water in the cell sap is reduced to zero. Since this is also the diffusion pressure deficit of the surrounding water, a dynamic equilibrium has been established due to the influence of the wall pressure upon the diffusion pressure of the water in the cell sap. When this condition of dynamic equilibrium has been attained, equal numbers of water molecules will be passing across the membrane in both directions per unit of time.

It is the turgor pressure of the cell acting in opposition to its wall pressure which imparts to plant cells their usual rigid, distended condition. This condition is termed turgor, turgidity, or turgescence; the first of these terms being in most general use. Plant cells, although generally turgid, seldom attain a turgor pressure equal to their osmotic pressure. Cells low or entirely lacking in turgor pressure are usually referred to as flaccid. Cell C will remain, according to the conditions prescribed for this experiment, in a state of incipient plasmolysis.

The diffusion pressure deficit of the water in the cell sap and in the solution are equal (*i.e.* their osmotic pressures are equal and neither is under any pressure, except, of course, atmospheric pressure) hence a dynamic equilibrium will be established as soon as the cell is immersed in the solution. Since there is no net movement of water into the cell no turgor pressure and hence no wall pressure will develop, a condition which will obtain in all plant cells in the condition of incipient plasmolysis. The diffusion pressure deficit of the water in the cell sap of cell B is 10 atmos., that of the water in the surrounding solution 6 atmos.

Hence the diffusion A the water in the solution is 4 atmos. greater than that of the water in the cell sap, and there will be a net inward movement of water into the cell. A turgor pressure thus develops within the cell, but at its maximum under these conditions it will attain a value of only 4 atmos. Counterbalancing this turgor pressure will be a wall pressure of 4 atmos. Since the initial diffusion pressure deficit of the water in the cell sap was 10 atmos., imposition of a wall pressure of 4 atmos. Reduces this to 6 atmos which is the diffusion pressure deficit of the water in the surrounding solution.

Hence the attainment of a turgor pressure of only 4 atmos. results under these conditions in an osmotic equilibrium. Since the magnitude of the turgor pressure developed in cell B will be less than in cell A, the latter cell will be more completely distended than the former. In all three of these cells a dynamic equilibrium has been attained by an adjustment of the diffusion pressure deficit of the water in the cell sap until it is equal to that of the water in the surrounding liquid.

This adjustment was attained in each of these cells by a shift in the magnitude of the wall pressure, which is one of the two components determining the diffusion pressure deficit of the water in the cell sap, the other being the osmotic pressure. Unlike a solution exposed to the atmosphere, the diffusion pressure deficit of a cell is not equal to its osmotic pressure except when the cell is flaccid, *i.e.*, possesses a zero wall pressure.

The term diffusion pressure deficit of a cell should be considered as an abbreviation for diffusion pressure deficit of the water in the cell sap. In the interests of simplicity the effects of such volume changes upon the osmotic pressure, etc. of the cell have been disregarded in the discussion up to this point. It is assumed that the wall pressure of the cell increases proportionately with an increase in the volume of the cell.

When the cell is completely flaccid its diffusion pressure deficit is equal to its osmotic pressure while its turgor pressure and hence its wall pressure are both zero. As the volume of the cell increases, due to an influx of water, the osmotic pressure decreases due to dilution of the cell sap. The turgor pressure and wall pressure increase equally and progressively with increase in the volume of the cell, the diffusion pressure deficit of the cell showing a corresponding but more rapid, progressive decrease.

When the cell attains a condition of maximum turgidity relative volume 1.5 its turgor pressure and wall pressure are equal to its osmotic pressure while its diffusion pressure deficit has fallen to zero.

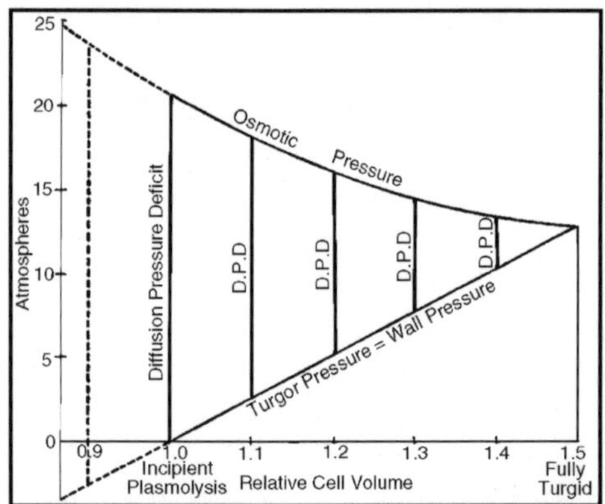

Fig. Inter Realtionships Among the Osmotic Pressure, Turgor Pressure, Wall Presure, Diffusion Pressure Deficit and Cell Volume of a Plant Cell.

The dotted extension of the line for wall pressure to the left is intended to suggest that it sometimes has a negative value. Conditions under which this may occur are discussed shortly. The diffusion pressure deficit of a cell in which the wall pressure is negative is greater than its osmotic pressure. In many types of cells volume changes are much less marked and hence they can often be disregarded in generalized considerations of the water relations of plant cells.

In some species many of the cells have walls which are totally lacking in elasticity. This is especially true of cells in the tissues of xerophytes and some water plants. The fundamental relations between the osmotic pressure, wall pressure, turgor pressure, and diffusion pressure deficit of a plant cell are expressed in the following simple equations: after attaining an equilibrium with a solution of 6 atmos. osmotic pressure, had an osmotic pressure of 10 atmos., and a wall pressure of 4 atmos. Hence its diffusion pressure deficit was 6 atmos and its turgor pressure was 4 atmos.

The quantity to which the name diffusion pressure deficit has been given in this discussion has been variously termed by writers in the field of plant physiology. Among the terms which have been used are suction, suction force, suction pressure, suction tension, water absorbing power, turgor deficit, and net osmotic pressure. All of the seterms are likely to be encountered in any extensive reading upon the subjectof the water relations of plants.

The authors of this book are not convinced that any of these terms is entirely free from criticism nor that any has as yet gained the sanction of even approximately universal usage. Hence we shall adhere to the term diffusion pressure deficit as a name for this physical quantity throughout this book. Those who have a decided preference for any other term need only to substitute it for diffusion pressure deficit wherever it occurs in the text.

CELL GROWTH IN ISOLATED FRAGMENTS

The elucidation of the mechanism of cell growth requires, of course, an analysis of the effects of different external conditions on the overall process and of each of its constituent aspects. Attempts have been made to develop such an analysis with intact organs.

The interpretation of the results obtained in this experimental situation, however, is a matter of considerable difficulty, since the effects observed may not be due to a direct influence on the growing cells themselves. In the intact root these metabolites may be supplied to the growing cell from mature tissue.

Clearly in this situation differences observed on growing cells may not be due to primary effects on these cells, but may represent secondary effects from influences exerted on mature or meristematic cells. This difficulty is avoided by studying the effects of different experimental treatments on isolated fragments composed of growing cells. This technique was first used by Bonner with *Avena* coleoptile tissue. This worker excised a 3 mm. fragment from the body of a coleoptile which had not reached the mature state, and subsequently observed the increase in length of the fragment in different media.

A similar technique has since been used by Christiansen and Thimann with fragments from pea internodes and by Brown and Sutcliffe with fragments excised from the extending zone of the root. This technique is an important one and has given valuable results. Nevertheless the limitations of it as it has hitherto been used require some consideration.

It is an important feature of the technique that the fragment is normally excised from an organ that has completed most of its development. Thus, although the mature coleoptile may be about 4 cm. long, the fragment is frequently taken from an organ which is about 2 cm. long, and the cells of which would therefore normally only grow to about double their size. In the overall growth of the coleoptile the increase in cell size must be considerably greater than this. Avery and Burkholder have shown with the *Avena* coleoptile from 4.0 to about 10 mm. growth is accompanied by cell division, and that from 1.O to 4.0 cm. growth is entirely by cell enlargement. Thus after cell division ceases there is at least a fourfold increase in length, and in view of observations on the root it is possible that cell size is increasing in this phase also, giving an overall increase in size greater than fourfold. A similar situation is involved with the pea stem and root fragments.

The root segments of Brown and Sutcliffe are taken from the zone which stretches from 1.5 to 3.0 mm. from the tip. The whole growth zone only occupies about 5.0 mm., and therefore at the time of excision at least the more mature cells in the fragment have completed most of their growth. The maturity of the tissue from which the fragments are taken no doubt accounts at least partly for the limited growth normally given under experimental conditions.

In favourable circumstances fragments from the coleoptile may double in length and those from the root quadruple. These increments, however, are considerably smaller than the normal twentyfold increase in volume with a tenfold increase in length that occurs when the cells are growing in an intact organ. The limited increases observed are of course highly significant with respect to the final stages.

On the other hand, in view of the results of observations on the changes in the enzyme complement during growth it is doubtful if results obtained with the final stages are applicable to the earlier. Since the metabolic pattern changes as the cell enlarges it is of course probable that the conditions that limit growth are different in the different phases of the process. This has been demonstrated by measuring the growth during a 24 hr. period of successive millimetre fragments excised from the root of the bean.

The fragments have been cultured with a 2 per cent sucrose medium on sintered glass disks. It increases with increasing distance from the tip until about 3.0 mm. and then decreases. Clearly while it is the sugar supply that limits growth in the 3-4 mm. fragment, it is not this factor which controls the situation in more immature tissue.

The growth after 4.0 mm. is less intense, since the cells are approaching the mature state. Thus the evidence suggests that observations with the excised fragments as they have been used hitherto are relevant only to an analysis of the final stages of growth, and it is within this limitation. Recent work has emphasized the importance of the metabolic state of the tissue in the determination of the course of cell growth.

Frequently, however, the rate and total amount of growth are considered as being linked to certain metabolic processes in the sense that an acceleration of the metabolic process leads to a corresponding acceleration in growth, and, conversely, that an inhibition of the metabolic process depresses growth. Evidence is assembled here, however, which suggests that growth may be interpreted in terms of two complementary sets of processes. One set when accelerated tends to promote growth, and a second when accelerated tends to depress it.

Whatever the metabolic processes involved are, however, it is clear that they can only affect the growth process through a comparatively limited set of physiological conditions, and stimulations or inhibitions of growth must be interpreted immediately in terms of these conditions.

The increase in volume of the fragment is determined of course primarily by the absorption of water which as in any other vacuolated cell is the resultant of the conditions tending to draw water into the cell and of the restraining pressure exerted by the wall.

Water may flow into the cell either along an osmotic gradient or as a result of a possible secretion into the vacuole which involves the consumption of metabolic energy. The importance of the osmotic gradient has been emphasized by Thimann and Schneider, who have shown that when coleoptile fragments are immersed in a solution of mannitol which is not absorbed growth is depressed.

In view of this it is evident that if growth is to proceed at a high rate the situation requires the continued absorption and accumulation of solutes in the vacuole, and the rate of solute absorption particularly when fragments are immersed in fluids of high osmotic pressure is clearly an important aspect of the situation. The restraining pressure of the wall in the earlier phases of the investigation of cell growth was defined in terms of elasticity and plasticity. These are no doubt important aspects of the situation.

It is evident, however, that the expansion of the wall and therefore the restraining pressure is determined primarily by the synthesis and deposition of wall constituents. Bonner showed with coleoptile fragments immersed in water or in fructose solutions that during growth there is a synthesis of wall material, and Brown and Sutcliffe found with root fragments that extensive growth only occurs when cellulose is being synthesized. At the same time it is probable that the synthesis of cellulose is only effective in promoting expansion when the physical state is such that the deposition of the cellulose in the wall can occur.

Gorter has recently shown that in root hairs which have been treated with triiodobenzoic acid conditions are induced which do not affect the synthesis of cellulose, but prevent the intussusception of it in the primary wall. Thus the growth of the fragment immediately depends on the operation of at least three and possibly four physiological conditions:

- A possible secretion mechanism for water,
- The absorption and accumulation of solutes,
- The physical state of the wall
- The synthesis of wall material.

These conditions are controlled by metabolic states, and at least some metabolic processes when accelerated modify the physiological conditions in such a way as to promote growth. Several workers have shown that a vigourous respiration is necessary for growth to occur. Bonner found with coleoptile fragments that extension did not occur in the absence of oxygen or in the presence of cyanide. Brown and Sutcliffe have shown with root fragments that growth is limited when the partial pressure of oxygen is reduced below that of air.

The promoting effect of a vigourous respiration is, however, a complex phenomenon, and in view of results considered below it is probable that it involves two aspects.

It is probable that it is a necessary condition for the promotion of certain other processes on which growth immediately depends, and that certain of the phases of respiration are directly linked to growth. That these two aspects are involved is suggested by a consideration of the effects of sugar and potassium on growth.

Thimann and Schneider reported that the growth of coleoptile fragments is enhanced by the provision of sugar, and Bentley has recently made a similar observation. Brown and Sutcliffe reported that in the absence of sugar the growth of root fragments is limited and that vigourous growth is only given in the presence of this nutrient. These workers analysed the sugar effect in terms of absorption and accumulation of sugar, synthesis of cellulose and respiration. They found that all these processes tend to increase with increasing concentration of sugar, but only when the cultures are exposed to air.

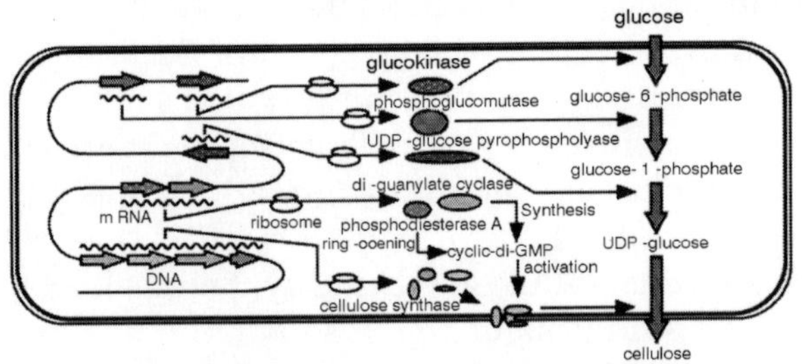

Fig. Cellulose Synthesis

In low partial pressures of oxygen they found a low respiration, a low absorption, no accumulation, and no cellulose synthesis. It was suggested that a high partial pressure of oxygen promotes a vigourous respiration which stimulates absorption. The rapid absorption in turn promotes the accumulation of sugar which establishes the osmotic gradient along which the absorption of water can occur. The rapid sugar absorption also provides the conditions for the synthesis of cellulose and thus stimulates the expansion of the wall.

This interpretation is consistent with the observation that there is no quantitative relation between respiration and growth. In this context respiration provides the necessary condition for the promotion of sugar accumulation and cellulose synthesis and is not directly linked to them. At the same time these are energy consuming processes, and a specific mechanism is presumably involved in effecting the energy transfer.

Bonner finds with coleoptile fragments that sodium arsenate inhibits growth but has little effect on respiration, and he suggests that this is due to the inhibition of the transfer of energy-rich phosphate bonds generated in respiration to reactions immediately involved in growth.

In the present connexion it may be suggested that respiration provides a source of energy-rich compounds which become involved in reactions immediately connected with sugar accumulation and cellulose synthesis. The effect of potassium suggests that there may also be a direct link between respiration and growth.

The influence of potassium in the growth of coleoptile tissue is uncertain; Thimann and Schneider found a stimulation, Frey-Wyssling an inhibition, and Bentley an inhibition. Whatever the position may be with coleoptile tissue, with root tissue it is clear and unequivocal. Brown and Sutcliffe found that in the presence of potassium growth and respiration are enhanced while cellulose synthesis and the absorption of sugar are not.

The enhanced growth cannot be attributed to a greater synthesis of cellulose. Also since the absorption of sugar is not increased when growth and consequent water absorption are raised, it follows that the enhanced growth occurs with a lower internal concentration of sugar. Brown and Sutcliffe suggest this might be due either to the adjustment of the internal osmotic pressure through an accumulation of potassium or to the stimulation of a secretion mechanism such as that proposed by Bennet-Clark, Greenwood and Barker.

It is significant, however, that potassium has no effect either on growth or respiration in the absence of sugar. Cartwright has recently examined the internal potassium concentration of the fragments used by Brown and Sutcliffe in the presence and absence of sugar and finds there is little difference. This suggests that the internal potassium concentration as such can have little effect on the osmotic gradient, and that the accelerating effect of potassium on growth is due to the stimulation of a water-secreting mechanism acting through a primary stimulation of some phase of the respiratory cycle.

The interpretation proposed here would involve an intimate connexion between some limited phase of respiration and the secretion of water in the sense that an inhibition of this phase alone would arrest the secretion of water. Commoner and Thimann have provided evidence for an intimate link of this general type between growth in the coleoptile and respiration.

They find that 10-5M-iodoacetate arrests growth but only depresses respiration by about 10 per cent, and they suggest that growth is controlled by this fraction of the total respiration to which it is linked. A vigourous respiration may also be involved in another connexion. During growth in the intact organ there is a net increase in protein content.

Even if the protein content does not increase, however, it is probable that protein synthesis must occur, and that this synthesis is necessary for the promotion of growth. The synthesis is an energy-requiring process, and can only occur when respiration is vigourous. The growth may be enhanced by the stimulation of certain metabolic processes. It is clear, however, from curves, that the overall growth process cannot be analysed in terms of such processes alone.

The effects of different concentrations of the same nutrient on the final lengths of the fragments suggest that growth is also determined by a complementary set of processes the stimulation of which tends to arrest it. The final length in 0.25 per cent sucrose is lower than it is in 2 per cent.

The lengths do not converge to the same limiting value at different rates depending on the external concentration. A similar situation is shown by the data of Schneider on the effect of different concentrations of sugar on the growth of coleoptile fragments.

If the stimulation of growth depended on the stimulation of a particular set of metabolic processes, with isolated fragments growth might be expected to continue until one or a complex of conditions became limiting. The limiting condition might be such as the protein level, the distance of the nucleus from the two ends of the cell, or some property involved in the geometry of the system.

Such a limiting condition, however, would be common to all members of an experimental series, and with different levels of a single nutrient they might all be expected to reach the same final length, but at rates which are directly related to the external nutrient concentration. Clearly some other factor is involved in the process, and this it may be suggested is the operation of an arresting mechanism which stops growth before a hypothetical maximum length can be reached in at least some series.

Other evidence is available which supports this conclusion. It has been shown that in the coleoptile the same concentration of heteroauxin is less effective if added 12 hr. after excision than it is if applied immediately. Similarly with root fragments, if these are transferred to water for 6 hr. and then to a sugar solution, the final length is less than it is if the sugar is supplied immediately. Clearly in the interval during which the stimulant has not been available changes have occurred which tend to restrict subsequent growth.

Secondly, there is evidence that with certain low partial pressures of oxygen the capacity for growth is retained longer than it is in air. When the oxygen level is very low the conditions of anaerobiosis have the normal effect of damaging the tissue. In 10 per cent oxygen, however, the arresting mechanism tends to be inhibited. In a series of experiments conducted by Sutcliffe fragments of maize roots were cultured in air and in an atmosphere containing 10 per cent oxygen, and at intervals during the experimental period cultures were transferred from the lower to the higher partial pressure.

With regard to continuous exposure to air and to an atmosphere containing 10 per cent oxygen, it is evident that while the rate of growth is lower in the lower partial pressure the time during which growth proceeds is very much greater. Clearly in the lower partial pressure the processes which promote growth are depressed giving a lower rate, while those tending to arrest growth are inhibited giving a longer period of growth.

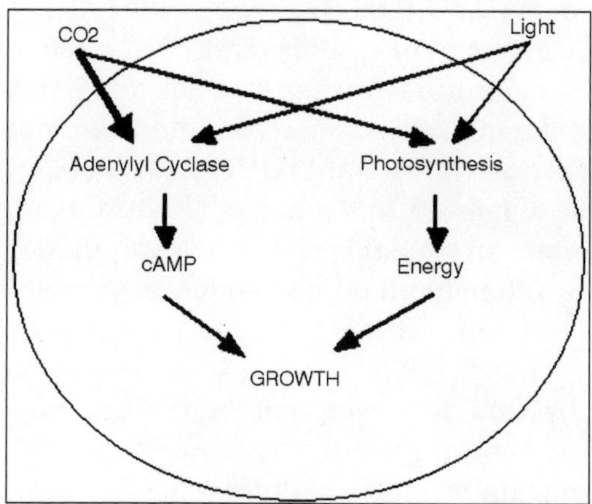

Fig. Various Metabolic Processes

The effects of transferring from the lower to the higher partial pressure are consistent with this interpretation. After transference the rate is accelerated, but this occurs after a period during which, although some growth has occurred, the growth-arresting processes have been inhibited. The net result is that the final lengths of the fragments are considerably greater than they would have been if they had been continuously in air.

Thirdly, it has been shown with root fragments that before growth ceases there may be a decrease in respiration. In intact roots it is probable that there are decreases in protein, the activities of certain enzymes, and respiration, either before or at about the time when growth ceases. The evidence suggests that progressive changes continue during growth, which when they have reached a certain critical level depress the rates of various metabolic processes, and it is as a result of such decreases that growth ceases.

The data obtained with isolated fragments which suggest the operation of an arresting mechanism do not indicate the nature of that mechanism. It may be suggested tentatively, however, that this involves the progressive inactivation of an enzyme system, in which case growth may be interpreted in terms of the rate per unit quantity of an enzyme or enzyme complex and the simultaneous absolute or relative reduction in the total quantity of this complex during growth.

In this situation it may be taken that the rate of the reaction per unit quantity of enzyme remains constant. The curve represents the change in the content of enzyme with time. Clearly the distance from this curve to the abscissa represents the rate of growth, and the area under the curve the total amount of growth made. It is immediately evident that at any one time the growth rate is higher, the total amount of growth made is greater and growth proceeds for a longer time. The lengths of the vertical lines here represent the rate per unit quantity of enzyme.

It is evident that when the total amount and the rate of growth are affected by the rate per unit quantity of enzyme and by the rate of reduction of the total quantity, then the only factor that can affect the time of growth is the rate of reduction of the quantity of enzyme, and if the time of growth is not affected then no effect on the rate of inactivation can be involved.

Thus it may be suggested that since potassium increases rate and final length but has no effect on time this is due to a stimulation of the promoting mechanism. On the other hand, sugar which undoubtedly stimulates the promoting mechanism, since it increases the time of growth, probably also inhibits the arresting mechanism.

It may be emphasized, however, that the present interpretation suggests that an enhanced growth in terms of rate and time could occur by a simple inhibition of the arresting mechanism only. It has been assumed that the rate of inactivation increases with time. It is evident, however, that the general character of the interpretation is not affected by the form of the inactivation curve. The arresting mechanism has been considered above in terms of a reduction in a total quantity of enzyme involved in growth.

0This is suggested by the changes in certain metabolic processes (notably respiration) that occur during the course of growth. Further, measurements have been made of the changes in the activities of certain enzymes, and it has been found with root fragments in certain instances that there may be a decrease in invertase activity and an increase in acid phosphatase activity. Thus changes during growth do occur in the activities of particular enzymes. These changes may be due to the production of specific inhibitors or to a transformation of one enzyme into another, in either case leading to a reduction in the total quantity of active enzyme. Another possibility may, however, also be considered. It is that growth arrest may be due to a relative and not an absolute decrease of a particular enzyme complex.

If two enzymes increase that compete for the same substrate the reaction catalysed by one enzyme might be virtually suppressed by a very much greater relative increase in the second. It is possible that the inactivation is of this type in the intact root, where there is a considerable increase in total protein during growth. On the other hand, an absolute decrease may be involved in the arrest of the growth of the fragments.

It was suggested above that the growth of fragments must be determined by four primary physiological conditions which in turn are regulated by metabolic states. It is clear, however, that the states that arrest growth must operate through these conditions no less than do those which promote it. The rate of growth of the root fragments used by Brown and Sutcliffe decreases with time. This decrease, is an expression of the increasing effect of the arresting mechanism, and should therefore be reflected in a corresponding change in one or more of the primary physiological conditions.

Brown and Sutcliffe, however, found that during growth of the fragments there are no changes in the rates of sugar absorption, sugar accumulation and cellulose synthesis. It is probable, therefore, that the effect of the arresting mechanism is expressed either through a change in the physical state of the wall or in the water-secretion mechanism, the effect of which might of course be determined by a change in the permeability of the protoplast to water.

4

Plant and Molecular Cytogenetics

Soon after the rediscovery of Mendel's Laws in the year 1900, the formulation of 'chromosome theory of inheritance' laid the foundation of cytogenetics, which has thus completed its first hundred glorious years in the year 2003. Subsequently, in the last few decades of the last century, the interest in plant cytogenetics that involved study and use of structural and numerical changes of chromosomes had largely declined. However, more recently, with interest in construction of molecular maps of chromosomes, and in the sequencing of whole genomes comprising an entire set of chromosomes, plant cytogenetics perhaps had a re-birth.

Other areas of recent research in plant cytogenetics include study of the organization of individual chromosomes in the form of chromosome territories and location of replication and transcription territories within the nucleus.

Chromosome structure has also been studied in great detail, so that the chromatin is now known to consist of nucleosome and chromatosome subunits, which undergo several levels of folding to contain it within the boundaries of the chromosomes. The structure of these nucleosome subunits has been studied at atomic resolution, and the structural relationship between adjacent nucleosomes has been recently resolved through X-ray crystallography of tetanucleosomes.

The structure of telomeres and centromeres have also been studied in several plant systems (particularly the cereals) in great detail suggesting that while structure of telomeres is conserved, but that of centromeres exhibit sufficient diversity.

In recent years, it has also been shown that both DNA and histone proteins, as components of chromosomes undergo large-scale modifications, which regulate gene expression. The study of these modifications also gave birth to the science of 'epigenetics and epigenomics'.

Projects like whole genome/epigenome sequencing of important plant systems are also yielding new information, giving new directions to research in plant cytogenetics. These exciting new developments in the field of plant cytogenetics will be briefly discussed in this discussion.

OVERLAPPING ERAS OF CYTOGENETICS RESEARCH

Progress of plant cytogenetics during the last ~80 years can be broadly divided in the following four eras, which may be slightly overlapping. Cytogenetics Era 1 (1910-1970): In this era chromosome number of a number of plant systems became known, structural changes like interachanges and inversions were studied for the first time in Stizolobium, Datura and Oentothera, inversions were studied in maize, and anuploids developed and cytogenetic maps constructed in several crops like maize, wheat, barley, etc. Alien addition and substitution lines were also developed in bread wheat using rye and few Aegilops/Agropyron species as the source of alien chromosomes.

Cytogenetics Era 2 (1950-1980): In this era, haploid DNA content (C-value) and composition (unique and repetitive) of nuclear DNA were determined in a large number of flowering plants using techniques of cytospectrophotometry and reassociation kinetics.

This led to recognition of two versions of C-value paradox. Firstly, the DNA contents in most eukaryotes were too high for the number of genes in the corresponding taxa, as estimated on the basis of known rates of mutations, and secondly the large-scale variation in DNA contents, could not be explained with the level of difference in complexity witnessed in these different organisms. The occurrence of large proportion of repetitive DNA in each of these eukaryotic genomes partly resolved the C-value paradox.

In this era, starting in early 1980s, DNA-based molecular markers were developed and molecular maps constructed in a large number of animals and plant systems, so that these molecular markers and the corresponding maps became an important resource for a variety of research problems, including their use in diagnostics and plant breeding. During this period, another significant development was the availability of a variety of fluorescence in situ hybridization (FISH), including multicolour FISH (McFISH), chromosome orientation FISH (CO-FISH), fibre-FISH, RNA-FISH, Comparative Genomic Hybridization (CGH) and 3-D FISH.

This era started in mid-1990s and gained momentum in the present century, with two distinct areas of cytogenetics research; first, the whole genome sequencing giving birth to 'reverse genetics', and second, the chromatin remodeling giving birth to the concept of 'histone code'. It is thus obvious that chromosome research has evolved and progressed at an incredible pace in the last two decades, more particularly during the past ten years. Since details of research done during Era 1, Era 2 and Era 3 would be widely available in text books and review articles, we mainly discuss the progress made during Era 4, in which significant progress has been made during the last few years to elucidate how the nucleosome and chromatin structure are modulated for expression of genes in time and space.

WHOLE GENOME SEQUENCING : GENOMICS AND EPIGENOMICS

In the year 2000, with the publication of the whole genome sequence of the crucifer weed, thale cress (Arabidopsis thaliana), plant cytogenetics entered into a new era of research, the era of plant genomics. Consequently in the early years of the present century, whole genome sequences of rice and a draft sequence of the genome of poplar (Populus trichocarpa) became available. Sequencing of several other plant genomes is also in progress (as listed at NCBI site)

According to some, the first wave of plant genome sequencing is over, and we are now entering a new era in plant genomics research. In this new era, genomes of many model species with small genomes or those of species of economic importance will be sequenced.

Also the available sequences will be subjected to annotation (assigning functions to these sequences), and the choice of new genomes to be sequenced will be made on several criteria, including phylogeny. The genomes of crops like maize and wheat will also be subjected to identification of gene-rich regions (GRRs), which will then be taken up for sequencing. There are also new technologies that will change the way we approach future genome sequencing projects.

DNA methylation, nucleosome remodeling (including histone modification and histone variants), and noncoding RNAs can organize chromatin into accessible ('euchromatic') and inaccessible ('heterochromatic') sub-domains. This extends the information potential of the genetic code, and one genome can generate many 'epigenomes' in time and space, during the life-span of an organism.

The implications of epigenetic research are far reaching, so that efforts are being made to study the epigenomes in a variety of eukaryotes including some plant systems. In a recent study, it was shown that these epigenetic modifications are not as conserved as was once thought. Further, very little is known about histone methylation in large genome plants, which make up the bulk of the angiosperms.

TRANSCRIPTION AND REPLICATION FACTORIES : A NEW CONCEPT

In recent years, specific regions have been identified within the eukaryotic nucleus, where replication and transcription takes place. It has also been shown that transcription sites are spatially distinct from replication sites. Therefore, we need to realize thattranscription by Pol II is not homogeneously distributed throughout the nucleoplasm, but occurs at highly enriched Pol II foci, known as transcription factories, which contain most of the hyperphosphorylated, elongating form of Pol II. Similarly replication takes place in 'replication factories'. It has been demonstrated that the chromosome segments destined to be transcribed or replicated have to move physically to these regions for transcription or replication to take place.

Transcription factories: A transcription factory is generally 80 nm tripartate structure mainly containing three spatially contiguous regions, with the template, the RNA polymerase II (Pol II) and the newly synthesised mRNA. The Pol II is perhaps attached to the transcription site, so that the template would move along RNAP II rather than the Pol II tracking along the template. Some translation also appears to be coupled with transcription at transcription factories.

There are fewer transcription factories than there are active genes and other transcription units in the nucleus, so that that more than one active gene is transcribed in each factory. And actively transcribed genes that are separated by long distances frequently co-localize in the same transcription factory. It has also been shown that actively transcribed genes co-localize with transcription factories, whereas identical, temporarily non-transcribed alleles, which can often be in the same cell, do not. Therefore, the 'on' state correlates with factory occupancy and the 'off' state with relocation away from factories. The specific Thus the specific nuclear repositioning of genes is correlated with transcriptional activation, silencing and replication timing.

The fact that different genes frequently co-occupy the same factory provides strong evidence that genes do not assemble their own transcription sites de novo when they become active, but instead migrate to preassembled transcription sites. A stable factory implies that genes or transcription units would essentially be pulled through a factory, rather than polymerases moving along the chromatin fibre, as is commonly believed. The finding that approximately 15% of the genome is transcribed — although probably not all at once — indicates that an extraordinarily large part of the genome passes through the limited number of transcription factories in a cell nucleus. This must have a profound effect on the nuclear organization of the genome.

Replication factories: It has been shown that a segment of DNA that needs to be replicated temporarily disengages Pol II and ceases transcription, although transcription in other regions of the genome continues uninterrupted throughout S phase. This segment of DNA needs to relocate itself in a replication factory. There seems to be no overlap between replication and transcription sites/factories.

CHROMATIN REMODELING

The term "chromatin-remodeling" generally refers to changes in histone-DNA interactions in nucleosomes. Histone proteins are modified and non-histone chromosomal proteins (*e.g.* high mobility group nuclear proteins = HMGN proteins; heterochromatin protein = HP1) are indirectly involved in the modification and activity of chromatin. Histone-DNA interactions in the nucleosomes are also modulated (remodeled) to facilitate interaction of other factors (not involved in remodeling) with DNA template. It is believed that the histone octamers of nucleosomes are often displaced from the enhancer and promoter regions by chromatin remodeling complexes to allow access of a variety of factors to DNA. During 1990s, evidence for chromatin remodeling and cellular memory became available in yeast (Saccharomyces cerevisiae), fruitfly (Drosophila melanogaster), and mammals, through genetic and biochemical studies.However, later towards the end of the last century and in the early years of the present century, chromatin remodeling has been studied and factors associated with this phenomenon in plant systems like Arabidopsis.

Activities catalyzed by chromatin remodeling: Chromatin-remodeling factors can catalyze the following activities:

(i) mobilization and repositioning of nucleosomes,
(ii) transfer of a histone octamer from a nucleosome to a separate DNA template,
(iii) the facilitated access of nucleases (enzymes) to nucleosomal DNA,
(iv) creation of di-nucleosome-like structures from mono-nucleosomes,
(v) generation of superhelical torsion in DNA, (vi) disruption of histone-DNA contacts; and
(vi) assembly and disassembly of nucleosomes.

Components of chromatin remodeling complexes: Three major strategies are used for chromatin remodeling. Each strategy makes use of a different set of proteins, described as components of chromatin remodeling complexes. Following are the three classes of chromosome remodeling complexes/ components:

(i) ATP-dependent chromatin remodeling complex, which makes use of ATP for chromatin modification through mere histone-DNA interactions.
(ii) Histone modifying enzymes, which are used for post-translational modifications of histones (mainly acetylation and methylation); these modifications create signals that define the so-called 'histone code'.
(iii) Variants of the histones (H2A, H2B, H3, H4 and H), which are synthesized and incorporated into nucleosome subunits (in place of normal histones), bringing about chromatin remodeling. More recently, the extreme use of H3 histone variants also led to the formulation of 'H3 barcode hypothesis'.

ATP-dependent chromatin remodeling machines/complexes: ATP-dependent :chromatin remodeling complexes (CRCs); also called chromosome remodeling machines (CRM), consist of ATPase polypeptide subunits and also non-ATPase subunits. Although ATPases play an major important role in chromatin remodeling, non-ATPase components also have a role to play. The

ATPases can be quite diverse and not all ATPases are important for chromosome remodeling. These ATPases are classified into three superfamilies, each superfamily having several families. One of these families belongs to a specific class of ATPases, described as 'Snf2-like family of ATPases', which are relevant to chromosome remodeling and therefore will be discussed in some detail.

(a) ATPases as subunits of ATP- dependent complexes: Members of SNF2-like family of ATPases are classified in several subfamilies, depending upon which protein motifs outside the ATPase region they have. Atleast seven subfamilies are known, but only four subfamilies are important, which include the following:
   (i) Swi/Snf2 subunit of the SWI/SNF complex;
   (ii) ISWI (ISWI, hSNF2H, hSNF2L, yISW1, yISW2; h stands for human, and y stands for yeast);
   (iii) CHD1 (CHD1, Mi-2a/CHD3, Mi-2b/CJD4, Hrp1, Hrp3),
   (iv) INO80 (CSB, Rad 26, ERCC6).
(b) Non-ATPase subunits of ATP- dependent complexes: The ATP dependent chromatin remodeling complexes also have non-ATPase subunits, which may have the following functions:
   (i) enhance or regulate motor activity of ATPase subunits;
   (ii) mediate other specialized functions that are not related with chromatin remodeling– these may be recruited to promoters via interactions with sequence specific transcription factors (TFs). Rad54 is also an example, in which SNF2-like ATPase is programmed by another polypeptide. Rad54 and Rad51 (related with bacterial RecA protein) catalyze homologous strand pairing.

Chromosome remodeling through histone modifications: The 'histone code': In eukaryotes the fundamental unit of chromatin is the nucleosome, which is a protein octamer/DNA complex composed of 200 bp wrapped around a histone octamer that consists of two molecules each of four core histones,

H2A, H2B, H3 and H4. One molecule of linker histone H1 is also associated with each nucleosome. Crystal structure of the core particle of this nucleosome unit consisting of 146 bp of DNA wrapped around histone octamer was determined in late 1990s at 2.8 A resolution.

More recently, crystal structure of a tetranucleosome was also determined at 9 Å resolution to understand the manner of higher-order folding of the nucleosome sub-units.

It was also shown in several recent studies that amino-terminal tails of histone proteins are targets for a series of post-translational modifications (PTMs), including acetylation, phosphorylation, and methylation. These modifications regulate chromatin structure and gene expression. Multiple histone modifications in various combinations are thought to form a 'histone code', since these modifications extend the information capacity of the associated DNA.

For example, histone H3 and H4 acetylation is consistently associated with transcriptionally active euchromatin, while methylation can be associated with either active or inactive chromatin depending on the residue involved in methylation.

For instance, methylation at H3K4, H3K36 and H3K79 are hallmarks for active transcription, whereas methylation at H3K9, H3K27 and H4K20 are correlated with transcriptionally inert heterochromatin. Further, the lysines in histone N-terminal tails can be mono-methylated (me1), di-methylated (me2) or tri-methylated (me3), and each methylation state may have unique biological functions, increasing the potential complexity of the histone code.

Table. Histone modification combinations and their effect on transcription

Nature of histone Modification	*Effect on transcription*
Histone acetylation	Transcription activation
Histone methylation	
H3K4, H3K36 and H3K79	Transcription activation
H3K9, H3K27 and H4K20	Transcription silencing

Table. Various combinations of histone methylations and their effect on state of chromatin condensation

Degree of methylation	*State of chromatin*
H3K9me3, H3K27me1 and H4K20me3	Highly condensed heterochromatin
H3K9me1, H3K9me2 and H3K27me3	Lightly condensed heterochromatin

It has been shown that methylation of H3K9, H3K27 and H4K20 is involved in heterochromatinization, but the degree of methylation of each of these lysine residues determines the degree of heterochromatinization.

For instance, H3K9me3, H3K27me1 and H4K20me3 mark the most deeply stained regions while H3K9me1, H3K9me2 and H3K27me3 mark the less condensed heterochromatin. Several reports indicate that the monoand di-methylated forms of H3K9 and H3K27 are enriched in heterochromatin, although degree of methylation also depends on the genome size.

For instance, unlike in animals, in Arabidopsis, H3K27me3 is associated with euchromatin and H3K9me3 is extremely rare.

There is also evidence that H4K20 methylation is also present in Arabidopsis. It is not clear, however, whether there is a direct relationship between heterochromatin and histone methylation. However, in plants with large genomes including maize and barley and wheat, very little is known about histone methylation, which makes up the bulk of the angiosperms. In a recent study, quantitative distribution of mono-, di-, and trimethylation at

H3K9 and H3K27 was examined in maize on whole genome level using oachytene chromosomes. The data reveal that three marks (H3K9me1, H3K27me1, and H3K27me2) correlate with DAPI (DNA) staining, but that only H3K27me2 is specifically enriched in condensed areas. It was also observed that H3K9me2 is not abundant in heterochromatin, but is instead enriched between chromomeres along with H3K4me2. The study also reported that centromeres contain H3K9me2 and H3K9me3; that H3K27me3 occurs at several brightly focused euchromatic domains, and that H4K20 methylation is rare or absent.

The Histone Code Hypothesis states that chromatin-DNA interactions are guided by combinations of histone modifications. For example, phosphorylation of serine residues 10 and 28 on H3 is a marker for chromosomal condensation; similarly, phosphorylation of serine residue 10 and acetylation of residue 14 on H3 is a tell-tale sign of transcription.

Chromosome remodeling through histone variants: 'H3 barcode hypothesis': In addition to post-transcriptional modifications (PTM) of histones, a number of variants are known for histone proteins that are found in the core particle of the nucleosome.

Perhaps H4 is the only exception, which does not seem to have any variant, but H3 and H2A are the two major classes of histones, which exhibit higher variation relative to H2B histone. Each histone variants differs from its corresponding normal histone in only a few amino acid residues, and each histone variant plays an important role in chromatin remodeling.

In particular, the histone variants for H3 offer the best example to illustrate the role of histone variants in chromatin remodeling, so that a H3 barcode hypothesis has also been formulated recently to describe the role of histone H3 variants.

The histone variants are generally synthesized directly from genes encoding them, but may also result due to post-transcriptional modification of conventional histones.

The genes encoding histone variants can be broadly classified into replication dependent (RI), replication independent (RI) and tissue-specific (TS), so that at least some of these variants occur in a tissue specific or developmental stage-specific manner. These histone variants replace the normal histones, and can be incorporated into the chromatin at any time during the cell cycle, although the conventional nucleosomes are produced and assembled into nucleosomes only during the S-phase of the cell cycle.

NUCLEAR ARCHITECTURE AND CHROMOSOMES

Chromosome organization within the nucleus: In the past, it was believed that chromatin within the nucleus is a network, intertwined and randomly distributed within the space available in the nucleus. However, both in plants and animal systems, recent evidence has demonstrated that the nucleus is a highly compartmentalized structure. Chromosome territories and interchromatin compartment (CT-IC model): The chromatin within the nucleus is organized in the form of chromosome territories(CTs) and interchromatin compartments (IC), and hence the formulation of CT-IC model. While CTs contain individual chromosomes, IC contains macro molecular complexes that are needed for replication, transcription, splicing and repair.

Newer techniques combining 3-D-FISH and computer aided deconvolution techniques helped in resolving the following features of chromatin organization and behaviour

(i) in an interphase nucleus, each individual chromosome occupies a discrete space, called the 'chromosome territory' or CT and that there is little inter twining among chromosomes;
(ii) in interphase cells, each chromosome also interacts with the nuclear envelope through consistent contact points;
(iii) in interphase cells, each chromosome interacts with other chromosomes through its heterochromatic regions; and
(iv) in dividing cells, chromosome movements are also non-random. The above information regarding chromosome organization at the physical level has also been integrated with genetic and molecular data, to decipher the mechanisms of different nuclear processes, in which chromosomes are involved, more particularly the transcription and DNA replication.

Rabl organization and telomeres orientation: Chromosome segregation at anaphase results in the polarization of chromosomes because sister centromeres are pulled in opposite directions and the rest of the chromosome arms trail behind. In some instances, this anaphase arrangement of chromosomes persists into the following interphase; this is known as the Rabl organization, in which chromosomes have a preferentially polarized organization, with centromeres at one end of the nuclear envelope, called the apical side, and telomeres at the opposite end, called the basal side. The presence of the Rabl organization is known to vary greatly between species and among tissues or developmental stages of an organism. In plants, it is generally observed in species with bigger genomes like those of wheat, rye, barley, and oats, but not in species with smaller genomes like those of sorghum and rice. Maize with intermediate size of genome displayed neither entirely Rabl nor entirely random chromosome organization.

Bouquet and telomeres clustering: The bouquet is the clustering of chromosome ends on the nuclear envelope (NE) during meiotic prophase, coincident with the initiation of homologous chromosome synopsis. The bouquet has been extensively described in many species in all eukaryotic groups and has been proposed as an aid to presynaptic alignment of homologous chromosomes. The similarity of the bouquet to the Rabl conformation has long been noted. However, it is clear that the two are not the same, although similar functions (chromosome pairing, recombination initiation and SC formation) have been assigned to both of them.

Relation between Chromatin Structure/ organization and Transcription

Heterochromatin and euchromatin: We know that the chromosomes of eukaryotes consist of darkly stained heterochromatin and lightly stained euchromatin. According to the classical view, heterochromatin is transcriptionally inactive and euchromatin is active.

This view is changing now; following are some examples: (i) there is evidence that several repeats within heterochromatic centromeres are transcribed to produce siRNA; (ii) it has been show that in each of the five chromosomes of Arabidopsis, heterochromatin is largely confined to the pericentromeric regions, and mainly consists of 180-bp satellite repeats, and retrotransposons (mainly from Athila family);(iii) NORs, mainly consisting of repetitive DNA, carries thousands of kilobases of tandemly repeated ribosomal DNA encoding rRNA. It has now been conclusively proved that in eukaryotic chromatin, cytosine methylation in repetitive DNA and distinctive modifications of individual histone proteins are frequent and cause heterochromatin formation. There are at least three kinds of methylases causing cytosine methylation in DNA: (i) Dnmt-1 type DNA methylase adds methyl-residues to cytosines (CG or CNG?) on newly synthesized strand of a DNA duplex, by using information from the conserved parental DNA strand carrying met-C, thus making DNA methylation heritable; (ii) Dnmt-3 type DNA methylase brings about DNA methylation (CG or CNG?) de novo, so that unmethylated DNA becomes methylated on both strands; (iii) chromomethylases (unique to plants) in Arabidopsis and maize recognize CpNpG sequences and bring about cytosine methylation.

MOLECULAR CYTOGENETICS

Major genome research programmes are underway to isolate expressed genes of corn and to DNA sequence at least the gene-rich regions. We are developing a highly efficient method for mapping DNA sequences to chromosome and sub-chromosomal regions based on lines derived from crosses between oats and corn. Partial hybrids have been obtained that contain an entire set of oat chromosomes along with one corn chromosome; lines are available that have each chromosome of corn individually added to the oat genome.

These lines are now being used to generate derivatives with only a subregion of the corn chromosome. This genetic material as well as B73xMo17 derived lines will be used to learn more about the subgenomic structure of corn and to test whether the degree of heteroallelism within and between the subgenomes contribute to heterosis. A pathogenic bacteria, called E. coli O157:H7, can contaminate hamburger and cause serious hemoraging. The contamination largely comes from the feces of cattle. Strains of E. coli have been selected that will kill the O157:H7 strain.

Two genes appear to be involved and this project will test their efficacy in corn. Wild rice is a crop under domestication. The seed falls (shatters) from the panicle upon maturation and is not available to be harvested. Storms can cause major losses of wild rice yield at harvest.

The shattering trait is considered a domestication trait that has been modified years ago in most major crops; the trait still exists in wild rice varieties. We have discovered major genes for shattering and have molecular genetic markers that should allow more efficient breeding for shattering resistant varieties. Methionine is an essential amino acid required in the diets of non-ruminant animals. Poultry have high demands for methionine accounting for why synthetic methionine is added to poultry rations in the U.S. A line of corn was selected that is high in methionine. Several generations of crossing, backcrossing, and selfing has resulted in relatively homozygous lines potentially high in methionine,

The specific objectives are:

1. Develop efficient mapping system for corn DNA sequences.
2. Learn more about the subgenomic structure of corn and relate the information to productivity.
3. Manipulate corn to produce compounds toxic to an enterohemoraghic bacterium (E. coli O157:H7).
4. Provide molecular genetic methods to the breeding programme to enhance the selection of varieties of wild rice with improved seed shattering resistance.
5. Finalize the selection and release of high methionine lines of corn.

We have produced lines of oats with a single corn chromosome added. These addition lines are available for all 10 corn chromosomes, individually present in the oat background. These lines are being used extensively for mapping DNA sequences to chromosome.

The oat-corn addition lines are being irradiated to produce radiation hybrid (RH) lines that have only part of the corn chromosome present. These lines allow much more refined mapping of corn DNA sequences. Future research will involve continued development and characterization of the RHs, mapping ESTs, cytogenetic transmission studies of the various lines, and obtaining the transmission of the chromosome 10 addition and formation of RHs.

Molecular genetic analysis of corn has shown that its genome is composed of major duplicated regions suggesting an ancient allopolyploid origin.

The potential exists for many loci to not only be duplicated but to have different alleles at each locus. Maximum productivity could be the result of maximizing the number of alleles at these loci. We will screen several genetic markers in order to determine Mo17xB73 lines containing duplicate regions containing various combinations of the parental genotypes.

Yield tests will ultimately be performed on selected lines with maximum heteroallelism in regions known to possess major yield QTLs. Our molecular genetics research on wild rice has resulted in the first molecular marker map and the placement of QTLs for several traits on the map. A PCR based marker has been developed that tags the most major QTL with polymorphism between shattering and non-shattering types.

This QTL accounts for about 40% of the variation in seed shattering in wild rice. A marker-assisted selection programme will be initiated using the molecular genetic marker for the major QTL and others as possible. Escherichia coli O157:H7 is a pathogenic strain of E. coli found in the digestive tract of beef cattle and causes food borne illness through contamination of beef during the slaughtering process. The goal of this project is to target the pathogen at the feed source or farm level to prevent contamination by developing a transgenic maize line that expresses a protein (termed "colicin") toxic to this strain of E. coli.

Colicin (produced by the cea gene) degrades DNA, and an immunity protein (produced by the cei gene) is required to prevent non-specific DNA degradation. Thus both genes are required to accomplish our goal. A screening procedure was developed in our lab that provides a screen for corn with elevated levels of methionine in the endosperm. A line was discovered, called BSSS5, that is resistant to lysine-threonine inhibition as a whole kernel but is inhibited when only the embryo is cultured.

The endosperm in this case possesses sufficient methionine to prevent the inhibition. BSSS 53 was found to produce a zein-2 protein with 23% of the amino acid residues as methionine. We have attempted to backcross the high-methionine trait into three different genetic backgrounds (B73, Mo17, and A632) and have some lines almost ready to release.

Oat-corn addition lines have been developed for every corn chromosome. These strains allow - in a single experiment - the rapid mapping of corn DNA sequences to their respective chromosome.

These strains resulted from more than 100,000 oat x corn crosses and isolation of over 7300 immature embryos. We extended the available genetic backgrounds by recovering oat-corn addition lines with B73 or Mo17. Between the two inbreds, we have a complete addition line set; the B73 addition lines will be useful in assembling the corn genome.

The DOE Joint Genome Institute will sequence chromosome10 using our Mo17 chromosome 10 oat-maize addition line. We also have added to our collection of radiation hybrid lines for each chromosome. These lines allow the placement of corn genes to specific regions of the chromosome. Corn gene expression/silencing in the oat-corn addition lines is being tested using the

DNA chip technology. Out of 17,000 tested sequences, 890 were expressed in the chromosome 5 addition.

About half of these sequences are of known position and are either placed on chromosome 5 or in a duplicated region of the genome. We further pursued the possibility of introducing C4 photosynthesis to oat with these materials.

Addition lines with both chromosomes 6 and 9 (carrying genes for key C4 enzymes) were tested for an effect on CO2 compensation point. A statistically significant change occurred in compensation point but not an impressive one. We will next investigate a third chromosome that has an effect on leaf cell morphology (known as Kranz Anatomy) involved in photosynthetic efficiency.

NIR analyses of self-pollinations of a new high oil accession indicate that the kernels contain 20% oil. A QTL analysis is underway. There is considerable interest in this material as a source of biodiesel and ethanol.

Using the molecular genetic map we published for wild rice, a PCR-based marker system has been identified for use in marker assisted breeding for the seed shattering trait. Genetic tests indicate a close marker/trait linkage.

The recent outbreak of the pathogenic E. coli 0157:H7 on spinach emphasizes the importance of having control measures to minimize such occurrences. We are developing what might be considered a prototype approach of using corn feed to alter the microflora in the intestinal tract as a means to reduce the presence of this pathogenic bacterium at its source.

Whole corn plants were regenerated from appropriate calli and found to carry the colicin gene, express the RNA and protein, and have anti-microbial activity against E. coli O157:H7. An extensive bioinformatics analysis has been performed of 1100 DNA sequences involved in human cancers to determine if they are present in plants. P53 is a protein involved in more than 50% of human cancers. We found a p53 binding sequence in the promoter of the MAD1 (Mitosis Arrest Deficiency) gene in human; MAD1 acts as a mitotic checkpoint to ensure chromosome stability. A MAD1 homolog was found in maize. Laboratory analyses of knockout mutants of MAD1 in Arabidopsis are underway.

Crop productivity is at least half due to genetic advances. Continued understanding of genetics and the application of molecular genetics will allow continued progress in food production. In addition, additional health benefits and identification of unique biofuel resources will be forthcoming.

POLYTENE CHROMOSOMES ALLOW BANDING AT A VERY FINE LEVEL OF RESOLUTION

In specialized cells of some plant and animal species, chromosomes undergo many cycles of DNA replication without undergoing mitosis. The result is polytene chromosomes, which may contain as many as a thousand parallel copies of each chromosome, perfectly aligned along their entire length. Polytene chromosomes can be stained and viewed in interphase nuclei. The most common example is the polytene chromosomes in the salivary glands of *Drosophila.* Because interphase chromosomes are more extended than those in mitotic cells, loci can be identified on polytene chromosomes at a much finer level of resolution.

FLUORESCENCE IN-SITU HYBRIDIZATION

Fluorescence In-situ hybridization (FISH) uses fluorescence to highlight the site of specific DNA sequences. Trask *et al.* 1993 provide a detailed review of the technique and its applications. In essence, a DNA sequence is 'tagged' by incorporating fluorescently-labeled nucleotides into the DNA chain. The tagged DNA, referred to as a probe, is incubated with chromosomes on a slide. Wherever the chromosome contains the same sequence as the probe, the probe can base-pair with the chromosome. The site to which the probe base pairs will fluoresce in UV microscopy.

In-situ hybridization is a valuable technique for studying chromosomal distribution of specific nucleic acid sequences and to determine the location of highly repetitive DNA sequences. Changes in these highly repetitive sequences reflect the physical rearrangements in the genome over time in the evolution of the species. Highly repetitive DNA sequences represent between 20 and 90% of the DNA in higher plants. The rate of evolution of individual repetitive sequences varies considerably; some may be conserved and be present in many species within a taxonomic family. The original technique published by Gall and Pardue (1969) has undergone numerous modifications and fluorescent in-situ hybridization (FISH) has been applied to chromosome identification and diagnosis in human genetic studies.

As with Geimsa banding, chromosomes must be fixed before denaturation. It is necessary to denature both the DNA probe and the target chromosome to allow complementary pairing to take place.

Fixative: The chromosomes are fixed in a 3:1 methanol:acetic acid solution. Slides are prepared by the squash method and the cover glass is removed to allow the cells to dry. The squash method stabilizes the chromosomes, permits good penetration of the reagents and brings most of the chromosome into the same focal plane, and also increases the speed at which the hybridization takes place, by decreasing volume.

Denaturation : incubate at 72° C in 50-70% formamide and 2X SSC (NaCl) (formamide denatures DNA helices)

Addition of Probes: The probe molecules are pieces of DNA, selected to target specific sequences on the chromosome. The target sequences can be as small as 1kbp or larger than 15 kbp. Sequences can be taken from entire chromosomes to highlight large regions of the genome. Denatured probe is hybridized with fixed chromosomes under a coverslip overnight at 37° C in

50% formamide, 2X SSC and 10% detran sulfate. After incubation, unannealed probe is washed off and the slides are labelled with immuno-fluorescent tags (ie. antibodies to the labeled nucleotides carrying fluorescent tags) and an anti-fading agent.

In summary, the chromosomes have been exposed to methanol, acetic acid, flattening, drying, rehydration, heat, formamide, and high salt concentrations during the FISH procedure. Some degree of structural changes will have been produced in the chromosomes. High resolution microscopic equipment is required for FISH as well as image processing software to enhance and analyse the image.

FISH is being used to study chromosome banding patterns, the distribution of chromatin in interphase nuclei, and chromosome abnormalities. Greater resolution is possible in interphase chromosomes due to the dispersed state of the chromatin, so that small structural changes can be detected. The compaction of metaphase chromosomes sets the limit of detection at sequences separated by more than 1 Mbp. The limit in interphase chromosomes is a separation of 100 kbp.

Detection of Single Genes

When the probe is a single gene, only that gene is expected to light up in FISH. The hybridization of a probe for the *neu* proto-oncogene in hyman lymphocytes. The *neu* locus resides on chromosome 17. Since the cells contain diploid nuclei, two hybridizing spots are seen in each nucleus. The sensitivity of FISH is put into perspecive by considering the fact that each chromosome is a single DNA molecule. That is, each spot results from a few probe molecules hybridizing to different parts of a single DNA molecule. Looked at another way, we can recall that Avogadro's number (the number of molecules in a mole) is 6.02×10^{23} molecules/mole, we are detecting 1.66×10^{-24} moles of chromosomal DNA!

Chromosome Painting

It is possible to isolate DNA specifically from individual chromosomes. When total chromosomal DNA is used as a probe, it will base pair with all sequences on the chromosome from which it was derived, making the entire chromosome 'light up'. In other words, each sequence in the probe eventually encounters its complementary sequence on the chromosome and base-pairs with it.

CYTOGENETIC TECHNIQUES

CHROMOSOME BANDING

A. Local variations in chromatin coiling, and the consequent variations in chromatin density, result in distinct staining patterns that can be used to identify each chromosome. The discovery of techniques which produce banding patterns on chromosomes is one of the most significant developments in cytogenetics. These techniques told us that chromosomes have an identity and structure that is consistent in all cells within a species, supporting the chromosome theory of inheritance. Staining made it possible to construct the first chromosome maps, and to distinguish between chromosomes of many different species. Fixed chromosomes are not equivalent to chromosomes in vivo. Fixation does not appear to extract DNA from chromosomes but it does extract histones and nonhistone proteins in varying amounts. The fixative may not always have a quantitative, reproducible effect on these proteins. The net result is that when you look at fixed chromosomes, you are looking at chromosomes that are chemically and structurally different from chromosomes in living cells.

Some relevant definitions:

- band - part of a chromosome that is clearly distinguishable from its adjacent segments by appearing lighter or darker.
- chromomere - any part of a chromosome that stains uniquely compared most bands.
- chromatin - the DNA/protein complex making up chromosomes
- heterochromatin - chromatin that tends to take up large amounts of stain, thus staining darkly. Heterochromatin tends to be densely-coiled.
- euchromatin - chromatin that tends to take up small amounts of stain, thus staining lightly.
- constitutive heterochromatin - chromatin that remains densely-coiled through out interphase, in all tissues. Often found near centromeres, telomeres and nucleolar organizing regions (NORs).
- facultative heterochromatin - chromatin that is either densely-coiled or loosely coiled during interphase, depending of the developmental state or tissue.

Giemsa Staining Methods

Mitotic metaphase is the best stage for studying chromosome morphology. Giemsa stain is specific for constitutive heterochromatin adjacent to centromeres and telomeres.

Geimsa C-banding technique has been used to identify individual chromosomes in many species by showing the position of constitutive heterochromatin. Geimsa staining methods were first described in a paper by Purdue and Gall (1970). However, it is important to recognize that staining methods typically need to be fine-tuned for each species, to give the optimal number of bands.We will concentrate on the general treatments and the basis for their application in terms of what each treatment is doing to produce the banding pattern.

Steps of C-banding:

1. Roots are harvested, pretreated and fixed in 3:1 95% ethanol:glacial acetic acid for at least 24 h. The roots are softened in 45% acetic acid or in 0.5% aceto-carmine. Slides are prepared by the squash method and the cover glass is removed using dry-ice method. Chromosomes adhere to the slide surface.
   The acid pretreatment depurinates the DNA (ie. removes purine bases without cleaving the sugar-phosphate backbone. This helps to induce nicks in the DNA backbone, loostening any supercoiling.
2. Dehydration. Typically slides are placed in 95 to 100% ethanol for 1 hour.
3. Denaturation. Treatment with barium hydroxide for 5 to 15 min at elevated temperature 50-55° C to denature the DNA. Alkali treatment denatures the DNA. ie. breaks base pairing, resulting in ssDNA.
4. Renaturation. The slides are then washed with distilled water and transferred to incubation at 60° C in saline sodium-citrate solution SSC (NaCl) which presumably renatures the DNA. Incubation periods and temperature are variable.
   The incubation in SSC alters the composition of the chromatin (DNA- protein complex). The differential reannealing of the C-band and the non-C-band chromatin to extraction is the basis for the differential staining. The heterochromatin which is highly repetitive DNA renatures under these conditions while the middle repetitive DNA and unique DNA does not. The result is the C-banding pattern.
5. Staining. Slides are then stained with Geimsa stain and checked periodically to see how the stain is progressing. When the optimal staining has been achieved, the slides are rinsed in distilled water to remove the excess stain, air-dried, stored in xylene overnight, air dried again and the cover slip is mounted using Canada Balsam, etc.

Other Geimsa banding techniques include G banding which provides more detail than C-banding. G banding provides identification of almost every chromosome in a complement and structural variation can be detected.

G-banding is produced using Giemsa (=G) staining and usually pretreatment with a diluted trypsin solution, urea or protease.

The Human Chromosome Study Group has published a diagrammatic representation of chromosome bands observed with G-,Q-, and R-staining methods. The nature of these bands appears to be stronger chromosome condensation, but the bands could also be interpreted as the result of alteration of histones and other proteins of the chromosomes. G bands cannot be produced in plant chromosomes possibly because of the high degree of condensation in plant metaphase chromosomes. N-banding was originally developed to stain the nucleolar organiser region of the chromosomes, plant and mammals.

STRUCTURAL BASIS OF CHROMOSOME BANDING

The exact role of underlying structural, functional and molecular organization of the chromosome in the determination of bands is still not fully understood. C-bands represent constitutive heterochromatin. G-banding patterns on mitotic chromosomes correspond very closely to the chromomere patterns of meiotic chromosome bivalents at pachytene, that is, the chromatin that is densely-packed enough to see. Chromomeres appear to be the focus of chromatin condensation along the chromosome, and may be the sites where chromatin condensation is initiated. If this is true, G- bands may be the corresponding initiation sites for chromosome condensation during mitosis.

Chromomeres may be more apparent in meiosis because of the relative degree of condensation of the chromosomes. Meiotic chromosomes are very extended and the homologues are paired which enhances chromomere patterns. Mitotic chromosomes are overall more condensed, masking chromomeres and requiring banding pretreatment to reveal them.

On human mitotic prophase chromosomes, up to 2000 or more bands can be observed. The bands on these high-resolution chromosomes correspond to meiosis chromomeres. As the chromosomes condense in the prophase to metaphase transition, there is a progressive coalescence of bands such that each band and its component sub-bands retain the same relative location and staining intensity. The bands observed at metaphase are made up of a collection of smaller sub-bands. The amount of constitutive heterochromatin in each chromosome is characteristic but polymorphisms do occur. The banding patterns are consistent and reproducible. Generally, features of chromosomes do not vary from tissue to tissue or change during the course of development. The organization reflected in banding is not random. In other words, we see consistent bands because the chromosome coils exactly the same way every time.

Constitutive heterochromatin is naturally differentiated by darker appearance during interphase and prophase, but does not usually show up in metaphase. It takes special methods to produce the bands.

In plant mitotic chromosomes, C-banding has been produced, but not G-banding, although pachytene chromosomes of plant meiotic cells demonstrate chromomeres. One suggestion is that this is due to the much greater quantity of DNA per unit length in plant chromosomes compared to those in mammals. The chromomeres are too tightly packed to be resolved by G-banding. The reason is not clear as compaction seems to be roughly equivalent. The preparation or pretreatment may have to be adjusted.

PHYSIOLOGICAL AND MOLECULAR WHEAT BREEDING

Plant Breeding is art and science of improving the heredity of plants for benefit of mankind. In Evolutionary concept, plant breeding is merely a continuation of the natural evolution of the crop species, changing its course of direction in the benefit of greater use to mankind. This also be defined as science of selection-Plant breeding is essentially an election made by man of the best plants within a variable population as a potential cultivar. In other words plant breeding is a 'selection' made possible by the existence of variability. Selections become the earliest form of plant breeding.

History of Plant Breeding

Plant breeding started with sedimentary agriculture and domestication of the first agriculture plants, the cereals were chosen by the early man. They learned to look for superior plants to harvest.

The domestication was hastened by early practice of harvesting mutant plants with special traits. This forms the ancient type of plant breeding. Before Mendal's discovery there was some plant breeding include selection and hybridization experiments. But the plant breeding was only hastened after discovery of Mendal's law on pea, thus lead a new science "Genetics". Modern plant breeding is applied genetics but its scientific basis is broader and uses conceptual and technical tools, molecular biology, cytology, systemetics, physiology, pathology, entomology, chemistry, and statistics (biometrics) and has also developed own technology.

Plant Breeding efforts may be divided into different historical landmarks-

Prehistoric Plant Breeding-Domestication of Crops

This includes domestication of crops by ancient people. Domestication is also continued so far.

Pre-mendel Plant Breeding

Economic botany and great Columbian exchange- There were some experiments before Mendel on hybridization and selection. Some seed companies were also established based on success on selection. Another part discovery of North America by Columbus in 1492 triggered unprecedented transfer of plant resources, first from old world to the new world, or the New World to the old. This increased variability in total genetic resources.

Mendelian Genetics and the Green Revolution

Mendel's experiment stimulated research by many plant scientists dedicated in improving crop production (plant breeders) through plant breeding. The most famous contribution of Mendelian genetics was hybridization. There was remarkable improvement in three economically important crops that made the food deficit world into a food surplus world. This is called the green revolution. The first, development of hybrid maize, the second development of high yielding and input responsive "semi-dwarf wheat" (CIMMYT breeder N.E. Borlaug received Nobel prize for peace in 1970), the third is high yielding "short sutured rice" cultivars. Similarly the remarkable improvements were done in other crops like sorghum and alfalfa.

Molecular Genetics and Bio-revolution

Totipotency shown by plants gave rise to tissue culture techniques such as somatic hybridization, doubled haploid production, clonal propagation, and in-vitro selection. Intensive research in molecular genetics has led to development of recombinant DNA technology (popularly called Genetic Engineering). Advancement in Biotechnological techniques has opened many possibilities for breeding crops. Thus mendelian genetics allowed plant breeders to perform some genetic transformation in few crops, molecular genetics provides key not only the manipulation of the internal structure but also their "crafting" according to plan.

Physiological and Molecular Wheat Breeding

Plant breeding has traditionally applied a trial-and-error approach in which large numbers of crosses are made from many sources of parental germplasm. Progenies are evaluated for characters of direct economic interest (*e.g.*, grain yield and grain quality) in target environments. Good performing parental germplasm, crosses, and progenies are selected for further use or testing. In many programmes "breakthroughs" in improvement are made simply by finding superior sources of parental germplasm among the numerous sources tested.

This conceptually simple approach has been highly successful in many crop species and numerous breeding programmes. The approach has often succeeded in the absence of in-depth knowledge about the physiological basis for superior performance. In some crops such knowledge has been obtained by doing retrospective analyses of prior genetic gains. Breeders have not applied this knowledge to a significant extent as a guide to further improvements, but instead have taken any avenue of improvement that happens to arise from direct selection for yield and economic performance. However with increased population, there is need to increase yield further and breeding require more scientific approaches to handle the problem.

Genetic Basis of Physiological Traits

During the past two decades, molecular tools have aided tremendously in the identification, mapping, and isolation of genes in a wide range of crop species. The vast knowledge generated through the application of molecular markers has enabled scientists to analyze the plant genome and have better insight as to how genes and pathways controlling important biochemical and physiological parameters are regulated. Three areas of biotechnology have had significant impact: the application of molecular markers, tissue culture, and incorporation of genes via plant transformation. Molecular markers have enabled the identification of genes or genomic regions associated with the expression of qualitative and quantitative traits and made manipulating genomic regions feasible through marker assisted selection. Molecular marker applications have also helped us understand the physiological parameters controlling plant responses to biotic and abiotic stress or, more generally, those involved in plant development.

Molecular Wheat Breeding.

Molecular wheat breeding is application of biotechnological tools in wheat improvement such as gene transfor (genetic engineering) and marker assisted selection. Such changes aims to alter the physiological pathways through change in genetic structures. There are many successful examples of such kind.

5

Transgenic Crops

The global area planted with transgenics crops has risen from 1.7 million hectares in 1996 to 90 million hectares in 2005 (James, 2005). Out of the total area of the world under transgenic crops, three countries namely USA, Argentina and Canada, which cover more than 90 percent.

The major crops occupying about more than 80 per cent area are herbicide-tolerant soyabean, Bt corn, Herbicide tolerant canola, and herbicide tolerant cotton. The three varieties namely Mech 12, Mech 162 and Mech 184 of Bt cotton have been released in India for commercial cultivation. Several transgenic cultivars of major food crops, such as soyabean, maize, canola, potato and papaya have been commercially released incorporating genes for resistance to herbicides, insects, and viruses.

Transgenic crops under commercial production across the globe.

According to (Manjunath, 2005), 04 field crops, 03 vegetables, 01 fruit crop and 01 tobacco crops are presently produced commercially across the globe..

Till date no transgenic crop variety tolerant to abiotic stress has so far been reported to be released for cultivation.

However, 7 transgenic varieties exhibiting tolerance to a range of abiotic stresses underwent field testing in Bolvia (frost tolerant potato variety), China (cold tolerant tomato), Egypt (salt tolerant wheat variety), India (Moisture stress tolerant Brassica variety) and Thailand (salt tolerant and drought tolerant rice varieties.

Most of the R&D activities on development of abiotic stress tolerant crops are being carried out in six countries of the Asian region namely Bangladesh, China, India, Indonesia, Pakistan and Thailand.

AGRICULTURE AND PROSPECTS OF COLD TOLERANT GENES

Cold is an environmental factor that limits the geographical distribution and growing season of many plant species, and it often adversely affects crop quality and productivity. Most temperate plants can acquire tolerance to freezing temperature by a prior exposure to low nonfreezing temperature, a process known as cold acclimatization.

Plants of tropical and subtropical origins are sensitive to chilling temperature (0°C-100°C) and are incapable of cold acclimation. Many studies have suggested that cold regulated gene expression is critical in plants for both chilling tolerance a and cold acclimation.

Cold responsive genes encode a diverse array of proteins such as enzymes involved in respiration and metabolism of carbohydrates, lipids, phenylpropanoids and antioxidants: molecular chaperones, antifreeze proteins, and others with a presumed function in tolerance to dehydration caused by freezing. The change in the gene expression occur in plant during cold acclimatization a developmental process that results in increased tolerance. Since then, it has repeatedly been speculated that certain COR (cold regulated) genes might have role in freezing tolerance. To test this notion investigators have turned to isolating the characterizing genes that are expressed in response to low temperature.

These efforts have led to the identification of a number of genes such as the COR 15a KINI LTI 78 fad 7 genes of Arabidopsis thaliana. It is well known to farmers and scientists that low temperatures can kill plants. Low temperature stress is a major environmental factor that not only limits where crops can be grown but also reduces yields depending on the weather in a particular growing season.

In addition to exceptionally stressful years that cause significant yield reductions, less extreme stress almost certainly causes smaller losses over large areas to produce comparable yield reductions every year. Even in cases when freezing stress does not result in yield loses., it often results in crop quality reduction.

Each year, worldwide losses in crop production due to low temperature damage amount to approximately $ 2 billion. Some of the major losses include the 1995 early fall frosts in the US which caused losses of over $1 billion to corn and soyabeans. The occasional freezes in Florida have shifted the citrus belt further south, and California sustained $650M of damage in 1998 to the citrus crop due to a winter freeze.

The inherent cold hardiness of the crop determines in which agricultural areas it can be grown. Crops that are more resistant to freezing stress would allow some geographical regions to grow more profitable and productive crops with less environmental risks. However despite continued efforts, traditional breeding has had only limited success in imparting crop plants with better freezing tolerance due to very little was known about the mechanisms that regulate chilling and freezing tolerance. With the advent of molecular genetics and biotechnology, it is now possible to genetically engineer plants to be more tolerant to many environmental adversities, including low temperature.

COLD ACCLIMATIZATION AND SIGNAL TRANSDUCTION CASCADE

Plants vary greatly in their ability to with stand low temperature stress. As an adaptation strategy, most plants native to the temperature climates develop freezing tolerance after prior expose to cold, Non-freezing temperatures, a phenomenon called acclimatization. Many biochemical and physiological changes are known to occur during cold acclimation. With the onset of low temperature, putative temperature sensors at the cellular membrane generates stress signals which are transmitted and amplified through many steps that include Ca++ signaling and a stepwise kinase/ phosphatase chain reaction termed the kinase cascade.

The massage eventually reaches the nucleus and regulators of gene expression called transcription factors, which act as master switches to regulate the expression of groups of genes, resulting in the increase of proteins and other organic molecules that protect the cell from freezing damage. The plants response to cold acclimation is clearly complex and diverse, and the actual biochemical and physiological changes are still poorly understood at the molecular level.

TECHNOLOGY DEVELOPMENT TOWARDS COLD TOLERANCE

The cbf genes represent one of the most significant discoveries in the field of low temperature adaptation and signal transduction. All-important crops and few vegetables and tree species have contained this gene. The transgenic canola contain this gene are able to survive freezing temperature as much as 4-50C lower than the non transgenic controls (Jaglo *et al.* 2001). Tomato plants have also been successfully engineered using the CBF genes to achieve chilling tolerance.

Thus the CBF technology has a great potential for improving the cold and freezing tolerance in plants. There is strong evidence that the decrease in fatty acids saturation that generally occurs in temperate plants upon exposure to low temperature contributes the cold tolerance of these plants.

Ishizaki-Nishizawa *et al.* 1996 found that by introducing a broad-spectrum desaturase gene from a cyanobacterium into tobacco plants could increase the low temperature tolerance of the transgenic plants. Expression of a plant phosphatase (At PP2CA) in transgenic Arabidopsis thaliana can accelerate the development of cold acclimation and increase freezing tolerance.

It has also shown that transgenic plants expressing a constitutively active kinase NPK1, is more tolerant to chilling and other abiotic stresses. Other transcription factors, including ABI3 and SCOF-1, have been used to successfully increases the cold tolerance of transgenic plants. Transcription factor genes Abl3, abi3, cbf1, dreb1A, dreb1 and dreb2 from A. thaliana, osmolyte biosynthesis genes from Arthrobacter globiformis and afb (anti freeze protein) from A. thaliana are some other important genes which have effective role to cold tolerance in plants.

IMPORTANT CROPS DEVELOPED THROUGH GENETIC ENGINEERING

A number of transgenic plants developed so far through plant genetic engineering for different agronomic and qualitative traits are:

- (i) Herbicide tolerant transgenic plants have been engineered by using mutants EPSP (5-enolpyruvyl shikimate-3- phosphate) synthase enzyme *e.g.* soybean. Canola, cotton and tobacco.
- (ii) Transgenic for insect resistance have been developed introducing a gene from Bacillus thuringiensis coding a toxic protein (delta endotoxin) which inhibits insect growth. *e.g.* Bt cotton, Bt maize, Bt tobacco.
- (iii) Transgenics for male sterility (Barnase and Barstar in Brassica spp. for hybrid seed production).
- (iv) Stress resistance transgenics. *e.g.* chilling resistance in tobacco, a gene for glycerol-1- phosphate transferase from Arabidopsis.
- (v) Transgenics for disease resistance such as fungal diseases, bacterial diseases, viral diseases.
- (vi) Transgenics for food processing, Bruise resistance tomato (expresses antisense RNA against).

MAJOR COMPONENTS FOR THE DEVELOPMENT OF TRANSGENIC PLANTS

Despite significant advances over the past decade, development of efficient transformation methods can take many years. Effective regeneration and transformation systems are the prerequisite for successful genetic transformation. Genetic transformation technology relies on the conceptual framework and the technical approaches of plant tissue culture and molecular biology to develop commercial process and products which comprises mainly the nine major components:

- (i) The development of reliable tissue culture regeneration systems
- (ii) Preparation of gene construct and transformation with suitable vectors
- (iii) Efficient techniques of transformation for the introduction of genes into crop plants
- (iv) Recovery and multiplication of transgenic plants.
- (v) Molecular and genetic characterization transgenic plants for stable and efficient gene expression

(vi) Transfer of genes to elite cultivars by conventional breeding methods if required

(vii) Evaluation of transgenic plants for their effectiveness in alleviating the biotic or abiotic stresses and field performance

(viii) Bio-safety assessment including food and environmental safety

(ix) Commercialization of genetically engineered crop.

TRANSGENIC DEVELOPMENT

Advancement in plant biotechnology has led to the identification and isolation of a number /transcription factor(s) related to abiotic stress tolerance. Defence Research and Development Organisation has taken up the work on development of transgenics for cold tolerance through cloning and transformation of these genes into high value vegetable crops to cultivate them in higher altitude areas for the fresh requirement of armed forces ane local populace. At present the work is carried out to develop transgenics with the following genes in tomato, cucumber, pea and brinjal and cloning of cold tolerant genes from seabuckthorn plant.

Osmotin gene: Among the various environmental factors, availability is the major factor influencing growth, development and productivity of crop plants.

Accumulation of low molecular weight osmolytes such as praline, betaines and sugar alcohol is an important mechanism underlying adaptation to such stress factors. In addition, stress induced proteins whose functions, however, remain largely unknown and enzymes that toxify reactive oxygen species can also contribute to stress tolerance. Osmotin is a basic 24 KD protein that was originally identified in tobacco adapted to NaCl and desiccation.

The stress induced synthesis and accumulation of the osmotin protein is correlated with osmotic adjustment in tobacco cells. The synthesis of osmotin is induced by ABA that accumulates in response to osmotic stress and subsequently plays a pivotal role in osmotic adjustment.

It is known that expression of osmotin gene induces proline accumulation in unstressed and stressed plants and imparts tolerance to both salinity and drought stress. The gene has been transferred through Agrobacterium in tomato at DARL, Pithoragarh and transgenic plants are obtained and T3 plants are being tested in high altitude areas under controlled environments.

Mannitol-1 phosphate Dehydrogenase (mtlD): Mannitol-1 phosphate dehydrogenase (mtlD) is one of the gene encoding enzyme that synthesize osmoprotectants and enhance their expression in transgenic plants which leads to maintenance of osmotic potential during the stress period. The candidate gene has been isolated from E. coli K-12 strain and code for an enzyme called mannitrol-1 phosphate.Mannitol-1 phosphate then converted into mannitol by non specific phosphates. Accumulation of mannitol in different parts of a plant leads to tolerance at cellular level by adjustment of the cytosolic osmotic potential when the concentration of electrolytes is lower in the cytosol than in the vacuole.

OSISAP1 gene of rice and CBF1 gene of Arabidopsis: The expression of these genes changes in the membrane lipid composition, accumulation of compatible osmolytes or cessation of plant growth. Cloning and expression of these genes has led to the development of strong tolerance to freezing through transgenic approaches. Dehydration responsive element binding factor (DREB/CRB3): The gene is responsible for the expression of many cold tolerant genes during cold stress in plants.

Glyceraldehydes phosphate Acetyl Transferase (GPAT) gene: The gene is responsible for the un saturation of the fatty acids present in the plant cell wall, which give the cold tolerance to the plants during cold stress. Cold regulated (COR15A) gene: This gene encodes for polypeptides that decreases the incidence of freeze induced lamellar to hexagonal II phase transitions, which increases freezing tolerance to the plants. Desaturase (desC) gene: The expression of this gene increases the fatty acid unsaturation of the cell wall lipids giving strength to the cell wall of the plant during cold stress.

TRANSGENIC PLANT

Transgenic plants possess a gene or genes that have been transferred from a different species. Although DNA of another species can be integrated in a plant genome by natural processes, the term "transgenic plants" refers to plants created in a laboratory using recombinant DNA technology. The aim is to design plants with specific characteristics by artificial insertion of genes from other species or sometimes entirely different kingdoms. Varieties containing genes of two distinct plant species are frequently created by classical breeders who deliberately force hybridization between distinct plant species when carrying out interspecific or intergeneric wide crosses with the intention of developing disease resistant crop varieties. Classical plant breeders use a number of in vitro techniques such as protoplast fusion, embryo rescue or mutagenesis to generate diversity and produce plants that would not exist in nature.

Such traditional techniques have never been controversial, or been given wide publicity except among professional biologists, and have allowed crop breeders to develop varieties of basic food crop, wheat in particular, which resist devastating plant diseases such as rusts. Hope is one such wheat variety bred by E. S. McFadden with a gene from a wild grass. Hope saved American wheat growers from devastating stem rust outbreaks in the 1930s. Methods used in traditional breeding that generate plants with DNA from two species by non-recombinant methods are widely familiar to professional plant scientists, and serve important roles in securing a sustainable future for agriculture by protecting crops from pests and helping land and water to be used more efficiently.

Natural movement of genes between species, often called horizontal gene transfer or lateral gene transfer, can occur because of gene transfer mediated by natural processes. This natural gene movement between species has been widely detected during genetic investigation of various natural mobile genetic elements, such as transposons, and retrotransposons that naturally translocate to new sites in a genome, and often move to new species over an evolutionary time scale. There are many types of natural mobile DNAs, and they have been detected abundantly in food crops such as rice.

These various mobile genes play a major role in dynamic changes to chromosomes during evolution and have often been given whimsical names, such as Mariner, Hobo, Trans-Siberian Express (Transib), Osmar, Helitron, Sleeping Princess, MITE and MULE, to emphasize their mobile and transient behaviour. Genetically mobile DNA contitututes a major fraction of the DNA of many plants, and the natural dynamic changes to crop plant chromosomes caused by this natural transgenic DNA mimics many of the features of plant genetic engineering currently pursued in the laboratory, such as using transposons as a genetic tool, and molecular cloning.

There is new scientific literature about natural transgenic events in plants, through movement of natural mobile DNAs called MULEs between rice and Setaria millet. It is becoming clear that natural rearrangements of DNA and horizontal gene transfer play a pervasive role in natural evolution. Importantly many, if not most, flowering plants evolved by transgenesis-that is, the creation of natural interspecies hybrids in which chromosome sets from different plant species were added together. There is also the long and rich history of interspecies cross-breeding with traditional methods.

Production of transgenic plants in wide-crosses by plant breeders has been a vital aspect of conventional plant breeding for about a century. Without it, security of our food supply against losses caused by crop pests such as rusts and mildews would be severely compromised. The first historically recorded interspecies transgenic cereal hybrid was actually between wheat and rye. In the 20th century, the introduction of alien germplasm into common foods was repeatedly achieved by traditional crop breeders by artificially overcoming fertility barriers. Novel genetic rearrangements of plant chromosomes, such as insertion of large blocks of rye (Secale) genes into wheat chromosomes ('translocations'), has also been exploited widely for many decades.

By the late 1930s with the introduction of colchicine, perennial grasses were being hybridized with wheat with the aim of transferring disease resistance and perenniality into annual crops, and large-scale practical use of hybrids was well established, leading on to development of Triticosecale and other new transgenic cereal crops. In 1985 Plant Genetic Systems (Ghent, Belgium), founded by Marc Van Montagu and Jeff Schell, was the first company to develop genetically engineered (tobacco) plants with insect tolerance by expressing genes encoding for insecticidal proteins from Bacillus thuringiensis (Bt).

Genetically Engineered Plants

The intentional creation of transgenic plants by laboratory based recombinant DNA methods is more recent (from the mid-70s on) and has been a controversial development in the field of biotechnology opposed vigorously by many NGOs, and several governments, particularly within the European Community. These transgenic recombinant plants (biotech crops, modern transgenics) are transforming agriculture in those regions that have allowed farmers to adopt them, and the area sown to these crops has continued to grow globally in every years since their first introduction in 1996.

Transgenic recombinant plants are generated in a laboratory by adding one or more genes to a plant's genome,and the techniques frequently called transformation. Transformation is usually achieved using gold particle bombardment or through the process of Horizontal gene transfer using a soil bacterium, Agrobacterium tumefaciens, carrying an engineered plasmid vector, or carrier of selected extra genes. Transgenic recombinant plants are identified as a class of genetically modified organism(GMO); usually only transgenic plants created by direct DNA manipulation are given much attention in public discussions. Transgenic plants have been deliberately developed for a variety of reasons: longer shelf life, disease resistance, herbicide resistance, pest resistance, non-biological stress resistances, such as to drought or nitrogen starvation, and nutritional improvement. The first modern recombinant crop approved for sale in the US, in 1994, was the FlavrSavr tomato, which was intended to have a longer shelf life. The first conventional transgenic cereal created by scientific breeders was actually a hybrid between wheat and rye in 1876.

The first transgenic cereal may have been wheat, which itself is a natural transgenic plant derived from at least three different parenteral species. Genetically modified organisms were prior to the coming of the commercially viable crops as the FlavrSavr tomato, only strictly grown indoors (in laboratories). However, after the introduction of the Flavr Savr tomato, certain GMO-crops as GMO-soy and GMO-corn where in the USA being grown outdoors on large scales.

Commercial factors, especially high regulatory and research costs, have so far restricted modern transgenic crop varieties to major traded commodity crops, but recently R&D projects to enhance crops that are locally important in developing counties are being pursued, such as insect protected cow-pea for Africa and insect protected Brinjal eggplant for India. Transgenic plants have been used for bioremediation of contaminated soils. Mercury, selenium and organic pollutants such as polychlorinated biphenyls (PCBs) have been removed from soils by transgenic plants containing genes for bacterial enzymes.

Agricultural Impact of Transgenic Plants

Outcrossing of transgenic plants not only poses potential environmental risks, it may also trouble farmers and food producers. Many countries have different legislations for transgenic and conventional plants as well as the derived food and feed, and consumers demand the freedom of choice to buy GM-derived or conventional products. Therefore, farmers and producers must separate both production chains. This requires coexistence measures on the field level as well as traceability measures throughout the whole food and feed processing chain. Research projects such as Co-Extra, SIGMEA and Transcontainer investigate how farmers can avoid outcrossing and mixing of transgenic and non-transgenic crops, and how processors can ensure and verify the separation of both production chains.

RISK ASSESSMENT OF TRANSGENIC CROPS

Lack of scientific data, non-scientific partizan views, uncertainty of potential risks, and ignorance confound rational discussion concerning the release of GMO. The issue of releasing genetically modified plants (GMP) into the farming system has become particularly agitated by lobbyist groups in Europe despite widespread cultivation of such crops in North America and elsewhere. Scientists must realise that the general public are concerned that an uncautious approach to the manipulation and cultivation of transgenic crops may affect biodiversity and its sustainable utilization in the farming system, *e.g.* loss of variability and viability. People also want that their views about applications of biotechnology for improving agriculture are listened irrespective of their knowledge in the subject. Moreover, farmers are afraid that negative propaganda jeopardizes the public image of their products. Scientists and policy markers should not forget that people's acceptability is the most important component of the general public assessment of risk, which includes both uncertainty and negative consequences. This acceptability depends on cultural factors because people's views change according to time and location.

The process of risk assessment in agro-chemical consists of (i) hazard identification, (ii) exposure assessment, (iii) effect's management, (iv) risk characterization, and (v) risk management. However, transgenic crops may be able to invade (or colonize) and multiply in many habitats. Hence, this risk assessment of a genetically modified living organism (also known as GMLO) must consider other characteristics not included when assessing the release of non-living compounds to the environment, *e.g.* horizontal gene transfer between transgenic crops and wild related species. Scientific risk assessment of transgenic crops must be strictly performed and precautionary principles should be considered in the decision making process. In the industrialized world, this precautionary principle is a key component of the response to the unforeseen (and sometimes irreversible) human and environmental impact, which may occur by introducing into the system new advances ensuing from research and technology development. In Norway, an unique legislation advocates that "the production and use of GMO should be ethically and socially justifiable in accordance with the principle of sustainable development" as well as "safe to humans and to the environment". By applying this framework, marketing applications of GMO could be rejected if insufficient documentation regarding ecological and heath aspects was submitted by the producer.

What are the potential ecological risks associated with the release of GMP into the farming system? These are of course a very large number of potential risks, However, perhaps the two most important risks are:

(i) GMP establishes in semi-or natural habitats, and
(ii) inserted transgenes incorporate into other species, thereby affecting non-target organisms in farms or natural habitats.

Hierarchical test protocols have been proposed to assess the risks of releasing GMP. Such protocols require knowledge about evolutionary history, morphology, life-history characteristics, pollination or breeding system, gene-transfer likelihood, natural hybridization, recruitment and vegetative propagation of a chosen species. Likewise, producers should provide, to facilitate this risk assessment, additional information regarding biochemical, physiological, and morphological changes owing to inserted gene(s), along with a list and description of marker and reporter genes included in the transgenic plant. It would also be important to add details concerning when and in which plant tissues or organs will be expressed the modified function or phenotype. Nonetheless, people must also know that scientists assessing risks of transgenic crops may extrapolate the outcome or results from simple short-time experiments into complex long-term natural-or farming systems. Investigations about gene flow and competing ability of transgenic crops may be easily addressed through short-term experiments. However, the assessment of the environmental impact of GMP requires a long-term, expensive, holistic research. Computer modelling, which integrates knowledge about gene flow, competing ability, spread of transgenes to weedy species, and cultural practices in the farming system, may provide an alternative means for long-term risk assessment of releasing GMP into the environment.

Consumer concern about transgenic crops also focuses on their safety as food, especially if modifications could influence their metabolism or health. In this regard, transgenic plants without selectable markers, such as antibiotic resistance genes, are needed to convince GMP-sceptics of the advantages of genetic engineering for crop improvement. In this way, their criticism concerning the potential risks of transgenic crops could be overcome. For example, molecular or metabolic markers may provide a means to identify transgenic plants with desired trait(s). Of course, these alternative markers should be safe from an environmental and health perspective.

OUTLOOK

Within the next 10 or 20 years, five research areas may become very important for crop improvement: (i) apomixis to fix hybrid vigour, (ii) male sterility systems with transgenics for hybrid seed in self-pollinating crops, (iii) parthenocarpy for seedless vegetables and fruit trees, (iv) short-cycling for rapid improvement of forest and fruit trees, and (v) converting annual into perennial crops for sustainable agricultural systems. The development of perennial crops will be especially important to protect the soil from erosion. Plant biotechnology will play, of course, an important role in achieving research and development success in these areas.

Banning transgenic crops in the farming system will be foolish because the potential benefits are so great. Environmentalists should recall or re-read 'Silent Spring' by Rachel Carlson (1962). Whatever scientists do to develop crops that eliminate or reduce the utilization of polluting agro-chemicals in the farming systems must be welcome by farmers and consumers. For example, one interesting approach for developing resistant transgenic crops may be through the improvement of the plant's own defence system. Inducible and tissue specific promoters could assist in this endeavour.

Collective approval may lead to new partnerships, cooperation or joint ventures in research and development between scientists in the public and private sectors that will benefit farmers and consumers with profits and high quality products, respectively. Any potential risk in human development associated with biotechnology applications in agriculture will be easily resolved in a democratic society. The public need to choose between being safely self-regulated or to follow safety regulations as agreed by lawmakers after listening to the views of scientists, producers, and consumers.

The general public should see biotechnology as a safe tool for scientific crop improvement, because it helps in the fight against hunger and poverty. Therefore, research funding should be allocated accordingly to long-term plant breeding programmes, which include biotechnology as one of its tools. In this way, we may effectively face the serious challenge of feeding the rapidly growing world population in the next millennium.

POLYPLOIDY IN PLANTS

One of the remarkable features of living material is their ability to perpetuate themselves. However, the ever dynamic nature of the surrounding environment has imposed upon plants, much like other organisms, various evolutionary and selective bottlenecks necessitating the adoption of ways and means by organisms to keep their "race going". A concept which gains prominence in this regard is that of hybridization and has been an area of much fascination since the eighteenth century. Stebbins (1958) defined hybridization as the "crossing between individuals belonging to separate populations which possess different adaptive norms"(27). Polyploidy, a prime facilitator of speciation and evolution in plants and to a lesser extent in animals, is associated with intra and inter-specific hybridization(12). The purpose of this chapter is to give a broad overview of the phenomenon of polyploidy in its entirety in plants ranging right from a brief historical background to where it stands today. Given the importance of polyploidy in speciation this will be looked into a little more in detail compared to other aspects of polyploidy.

So what is polyploidy? It refers to a definite arithmetic relationship between the chromosome numbers of related organisms (2). It has been defined as the possession of three or more complete sets of chromosomes and has been an important feature of chromosome evolution in many eukaryotic taxa including plants, yeasts, insects, amphibians, reptiles, fishes and even the mammalian genome (19). Polyploidy is present to at least to some extent in most members of the plant kingdom being more common in some and rather rare in others(26). The fact that it is widespread many plants is also kind of exemplified by the wide variations in chromosome numbers with chromosome numbers ranging from 2n = 4 to 500 in angiosperms to 2n=6 to 226 in monocots(2). Thousands of angiosperm species have 14 to 15 pairs of chromosomes (2).

The phenomenon of polyploidy gained much of what it is today during the early part of the twentieth century. One of the early examples of a natural polyploid was one of De Vries's original mutations of *Oenothera lamarckiana* (mutat. *gigas*). The first example of an artificial polyploid was by Winkler (1916) who in fact introduced the term polyploidy. Winkler was working on vegetative grafts and chimeras of *Solanum nigrum* and found that callus regenerating from cut surfaces of stem explants were teratploid. Digby (1912) had discovered the occurrence of a fertile type *Primula kewensis* from a sterile inter-specific hybrid through chromosome doubling but failed to realize its significance in the context of polyploidy. Though unaware of the 'Primula type" fertile hybrid, Winge (1917), from his studies on the chromosomal counts of *Chenopodium* and *Chrysanthemum* found that chromosome numbers of related species were multiples of some common basic number; he subsequently proposed a hypothesis that chromosome doubling in sterile inter-specific hybrids is a means of converting them into fertile offsprings. This was subsequently verified by various workers in artificial inter-specific hybridizations of *Nicotiana, Raphanobrassica* and *Gaeleopsis*. Finally the colchicine method of chromosome doubling was developed by Blakeslee and Avery (1937) and became an important tool for the experimental study of polyploidy. Before going ahead what is the main significance of polyploidy in brief? As pointed out at the very outset it is recognized as one of the main process in the evolutionary history of plants and to some extent other organisms. Though differences do exist with regard to the nature of its role and the relative importance of different kinds of polyploidy, an understanding of the ways in which polyploidy operated in the past to produce new species and races may provide useful insights to improving our cultivated plants. In fact many of our crop species, including wheat, maize, sugar cane, coffee, cotton and tobacco, are polyploid, either through intentional hybridization and selective breeding (*e.g.* some blueberry cultivars) or as a result of a more ancient polyploidization event (*e.g.* maize).

Added to it, technological advances in the analysis of genome structure and function has made it possible to better analyze the genetic consequences of genome duplication. The importance of polyploidy in such diverse fields such as cytogenetics, physiology, breeding, cytotaxonomy and biogeography in conjunction with new possibilities put forth by various molecular techniques has all spurred a resurgence of interest in issues of origin and establishment of lineages.

Origin of Polyploids

There are various modes for the origin of polyploids. These mainly include mechanisms such as somatic doubling during mitosis, non-reduction in meiosis leading to the production of unreduced gametes, polyspermaty (fertilization of the egg my two male nuclei) and endoreplication(replication of the DNA but no cytokinesis). Endoreduplication however, is more similar to somatic doubling and is therefore not viewed as a separate mechanism by some authors.

Chromosome doubling can occur either in the zygote to produce a completely polyploid individual or locally in some apical meristems to give polyploid chimeras. Somatic polyploidy is seen in some non-meristematic plant tissues as well. (*e.g.*: tetraploid and octoploid cells in the cortex and pith *Vicia faba).* In somatic doubling the main cause is mitotic non-disjunction. This doubling may occur in purely vegetative tissues (as in root nodules of some leguminous plants) or at times in a branch that may produce flowers or in early embryos (and may therefore be carried further down). Spontaneous somatic chromosome doubling is a rare event and the only well documented instance of the same was in case of tetraploid *Primula kewensis* which arose by somatic doubling in certain flowering branches of a diploid hybrid. The phenomenon of chromosome doubling in the zygotes was best described from heat shock experiments in which young embryos were briefly exposed to high temperatures. Zygotic chromosome doubling was first proposed by Winge and the spontaneous appearance of tetraploids in *Oenothera lamarckiana* and amplidiploid hybrids in *Nicotiana* were shown to be a result of zygotic chromosome doubling. A second major route of polyploid formation involves gametic "non-reduction" or "meiotic nuclear restitution" during microsporogenesis and megasporogenesis resulting in unreduced 2n gametes. Non reduction could be due to meiotic non-disjunction (failure of the chromosome of separate and subsequent reduction in chromosome number), failure of cell wall formation or formation of gametes by mitosis instead of meiosis.The classic example, *Raphanobrassica,* originated by a one step process of fusion of two non-reduced gametes. The production of non-reduced gametes has been shown to be rather common in *Solanum sps*..

Another route may involve non–reduction occurring in one of the germ lines. A tetraploid individual can then result from a two-step process (sometimes referred to as a triploid bridge mechanism) from the fusion of an unreduced 2n gamete with a reduced 1n gamete to give a 3n zygote followed by the subsequent fusion of a 3n gamete with a normal 1n gamete in the next generation to give rise to a tetraploids individual (as in artificial *Galeopsis tetrahit*). This is a more common route of polyploid formation from unreduced gametes (though the frequent sterility of most triploid hybrids has lead to a questioning of this method by some authors rather than the former one. The production of non-reduced gametes is also a function of the environment and genotype. eg: adverse growing conditions were shown to favour an increase the number of non-reduced gametes in *Gilia*. An example of the influence of genotype in modulating the production of non-reduced gametes can be seen in case of maize wherein the gene *"elongate"* on chromosome 3 was found to increase the proportion of diploid eggs produced.

Studies on unreduced gametes in both plants and animals are getting easier with the use of rapid screening techniques such as flow cytometry, chromosme painting and other genomic techniques. Polyspermy is observed in many plants but its contribution as a mechanism for polyploid formation is rather rare except perhaps in some orchids.

Endoreduplication is a form of nuclear polyploidization resulting in multiple uniform copies of chromosomes. It has been known to occur in the endosperm and the cotyledons of developing seeds, leaves and stems of bolting plants. In animals it occurs in certain tissues such as the liver cells, and megakaryocytes (the cells which give rise to the thrombocytes).

Miscellaneous Factors Promoting Polyploidy

There are a number of other factors favouring polyploidy include (but not limited to), the mode of reproduction, the mode of fertilization, the breeding system present, the growth habit of the plant, size of chromosomes etc.

Polyploidy seems to be favoured in long lived/perennial plants possessing various vegetative means of propagation (eg: *Fragia, Rubus, Artemisia, Potamogeton* etc.) and in those with frequent occurrences of natural interspecific hybridizations. Various possible reasons have been advanced by various workers to account for the above phenomenon; one of the widely accepted ones' being the enhanced chances of somatic doubling made possible in plants with enhanced lifespan and vegetative means of reproduction..

Cross fertilization and allogamy were argued to be factors favouring polyploidy. Autogamy however was thought to restrict it. There have been opposing views as well on the same. For example in the tribe *Midinae* (Compositae) polyploidy was found to be highly developed in the autogamous genus *Madia* but almost absent in the allogamous species of *Layia* and *Hemizonia*.

As regards latitude and altitude, the proportion of polyploids has been found to increase with latitude and altitude; but this has not always been found to be true with respect to altitude. Various reasons which have been put forth to explain the above trend in the distribution of polyploids some of them being the better adaptability of polyploids to colder climates and changes that might have taken place in the Pleistocene period etc. Various ecological factors also have a bearing on the distribution of polyploids eg: polyploids were found to be more frequently distributed in wet soils and meadows as opposed to more stable habitats with drier soils or forest communities respectively. With regard to the breeding system since the main mode of origin of allopolyploids in annuals is by the fusion of unreduced gametes, the presence of an outcrossing breeding system tends to reduce the chances of union of unreduced gametes.

6

Biotechnology of Plant Breeding

Biotechnology is already becoming a powerful presence in our lives, yet most of us know very little about this important scientific phenomenon. Some of us are somewhat put off by technology, and sometimes frightened by it, but we can't keep ignoring it. The purpose of this chapter is to explain in simple terms what biotechnology is and to illustrate some of the positive aspects of biotechnology that are already benefitting, or are expected to benefit, the New Zealand fruitgrower.

How does this technology work? All living things carry their genetic inheritance in chemically coded messages or DNA (deoxyribonucleic acid). Under the microscope, DNA looks like a mass of tangled threads. But these threads consist of tiny subunits of DNA called genes. Genes carry instructions, sometimes called "the blueprint of life", for things such as fruit size or skin colour. Genes programme these instructions through arrangements of just four simple molecules-adenine, cytosine, thymine, and guanine. These A, C, T, and G's are like letters of the alphabet-their order "spells" out the language of life.

In nature genes (*i.e.* pieces of DNA) sometimes "mutate", *i.e.*, alter their chemical structure spontaneously from time to time. These mutations may result in plant cultivars carrying valuable new traits. For hundreds of years, farmers have taken advantage of natural mutations to improve traditional plant lines. For example, the popular apple cultivar "Royal Gala" is a natural mutation of "Gala". However, the time taken from a chance natural mutation to the production of a valuable new cultivar can be many years. But with genetic engineering, scientists are now able to shorten this time considerably.

The term "genetic engineering", sometimes also called "genetic manipulation" or "recombinant DNA", describes how scientists insert new genes into a plant and change its genetic makeup. This process is something like tearing out a page of the instruction manual for one living thing and glueing it into the manual of another. The resulting plant which contains a new gene is called "transgenic". In the past, we could breed using only closely related species and it took many generations to introduce significant improvements. Genetic engineering allows researchers to shorten this time considerably by inserting only the gene responsible for a desired trait (*e.g.*, pest or disease resistance, skin colour, or fruit size) into a plant, and selecting for that trait. This mimics the evolutionary process but in a highly selective way. Biotechnology here refers to the commercial application of genetic engineering to the development of new plants. Biotechnology could play a key role in improving our food supply. It could provide us with more nutritious crops and longer-lasting food products that are easier to store and handle. Many genetically engineered plants are already growing in test fields overseas: canola oil with a healthier fat profile; a tomato with longer shelf life and better flavour; alfalfa and cotton with resistance to powerful herbicides; and cotton with resistance to insect pests.

INTRODUCTION TO PLANT BREEDING

IMPROVING CROP PERFORMANCE

The dramatic gains in agricultural productivity seen in the second half of the last century are often linked uniquely to increased mechanisation and the widespread introduction of fertilizers and agrochemicals.This 'Green Revolution' would not have been possible without the huge genetic improvements made to the crops. Plant breeding has contributed around half of the threefold increase in UK wheat yields recorded from 1947 to 1992 and yields have continued to improve since then. Plant breeding can directly improve the performance of crops in different ways.

Productive Yield

Developing crop varieties, which convert more of their biomass into productive yield, is the single biggest contributor to improved crop output. The introduction of shorter-strawed cereals is a striking example of how this has been achieved, by transforming more of the crop's productive energy into valuable grain.

Physical Characteristics

Changing a crop's physical structure can also contribute to increased yields. For example, the development of semi-leafless varieties of field peas helped to boost intrinsic yield. It also stimulated the growth of stabilising tendrils, which significantly improved the crop's standing ability so reducing crop losses at harvest.

Disease Resistance

Genetic resistance to disease enables crops to realise their yield potential–it can also mean reduced use of agrochemicals. Plant breeding has significantly improved the genetic resistance of crops against the threat of viral and fungal infection. Key examples include resistance to blight in potatoes, rhizomania in sugar beet, and barley yellow mosaic virus in cereals. The challenge of breeding resistant varieties is constant because new strains of disease develop naturally.

Timing of Maturity

Plant breeding technology has brought major improvements in the uniformity with which crops ripen ready for harvest. This not only reduces potential crop losses at harvest (as in the case of pod shatter in oilseed rape), but has also improved growers' ability to mechanise harvesting operations. In the field vegetable sector, for example, the labour intensive (and unpleasant!) task of picking crops such as cauliflower, broccoli and Brussels sprouts has been transformed by the development of improved varieties.

Other Agronomic Factors

By improving crops' ability to cope with a range of other agronomic pressures, advances in plant breeding continue to underpin progress in agricultural productivity.

They include:

- Genetic resistance to pests, such as nematode resistance in potatoes
- Shorter crop life-cycle, to expand a crop's geographical growing areas
- Stress tolerance, such as frostresistance in field vegetables, to extend the seasonal availability of home-grown fresh produce.

Improving Crop Quality

Alongside progress in keeping Britain's farmers competitive through improved productivity, developments in plant breeding have also brought significant gains in crop quality. Gone are the days when Europe encouraged farmers to concentrate solely on maximising output, assured that their crops would find support under the Common Agricultural Policy.

Today, as production-based support and barriers to international trade continue to be reduced, Britain's growers must compete with the best in the world, on quality and cost of production. Consumer expectations of food are also more exacting. This is reflected in tighter quality specifications along the food chain. Plant breeders have responded with a continuous stream of new varieties, tailored to the needs of specific end-markets.

Cereals

Britain's historical dependence on North American imports of wheat for bread making has been reversed over the past 40 years thanks to improvements in baking technology and the development of high protein, hard milling varieties of wheat.

Until the mid-1960s, imported wheat typically accounted for up to 65% of our bread making requirements. Today as much as 90% of our bread is produced from homegrown varieties. In barley, improvements in malting quality have underpinned a fourfold increase in the volume of beer brewed per tonne, from around 2,000 liters in 1950 to nearly 8,000 liters today.

The quality of malting barley varieties grown in Britain is internationally recognized, with export markets from the near Continent to the Far East.

Oilseed Rape

Oilseed rape has become a major homegrown source of vegetable oil and animal feed within the past 30 years. The rapid expansion of this important break crop from 50,000 hectares in the mid-1970s to around 500,000 hectares today can be linked to two major breakthroughs in plant breeding. Varieties for food use were first developed in 1970. These contain reduced levels of erucic acid in the oil, making it more suitable for human consumption. This was followed in the late 1980s by 'doublelow' varieties, which also offered reduced levels of glucosinolates in the residual meal to improve quality for animal feed.

Potatoes

In the potato sector, plant breeders have responded to increasing consumer demands for quality and choice. Fresh produce counters now boast a bigger range than ever, from traditional main-crop varieties through to baby new, baking and salad potatoes. Breeding programmes have also targeted processing and catering markets, with specific varieties used to produce crisps, instant mash and chips.

Sugar Beet

Higher root yield and increased sugar content have combined to double sugar production per hectare in the past 50 years. The introduction of monogerm seed, allowing seeds to be planted individually rather than in clusters, ensures genetic improvements are fully realised in the field.

Pulses

As interest in home-grown sources of protein has increased, breeders have developed varieties of peas and beans for specialist markets–freezing and canning for human consumption, flaking for pet food, and tannin-free varieties for animal feed.

Creating a New Variety

The creation of each new variety is a complex, costly and skilled operation. It is also painstakingly slow–today's breeding programmes are already looking ten years ahead to the needs of farmers, consumers and the environment at the end of the next decade and beyond. Techniques vary between crop species, but the scientific principles of plant breeding remain unchanged from Mendel's first discovery that selected parent plants can be cross-pollinated to combine desired characteristics. Genes–units of hereditary material that are transferred from one generation to the next, determine plant characteristics. Since each plant contains many thousands of genes, and the breeder is seeking to combine a range of traits in one plant, such as high yield, quality and resistance to disease, developing a successful variety is an extremely lengthy process–up to 12 years in the case of cereals, even longer for potatoes. Plant breeding has been compared to playing a fruit machine–not with three reels, but several hundred. The skill of the plant breeder lies in improving his chances of hitting the jackpot by combining all the desired characteristics in the same variety.

RESPONSES TO BIOTECHNOLOGY IN CROP IMPROVEMENT

The advances in plant transgenics and genomics described above have not been isolated from society. Some of these achievements have been acclaimed by end-users whereas other accomplishments, *e.g.* release of genetically modified organisms (GMO), are being attacked, not only in words but also in deeds, by political activists.

Some of these educated middle-class campaigners are expressing in this way their rampant 'eco-paranoia', while others hide their real agenda to manipulate the fashionable ecological movement. This controversy has attracted the attention of non-scientific partizans to each side. There have been negative comments about transgenic plants by a crown prince and contrasting positive comments by a former president, both of whom may not have the required technical knowledge to assess the potential of biotechnology for crop improvement. Irrespective of this ideological dispute and ensuing democratic disagreements, biotechnology products will be accepted by people who support scientific-based progress, in a similar way that new cultivars or innovative crop husbandry techniques have previously become integral parts of farming systems elsewhere. However, without end-user's consent, the impact of a new technology in the society will be small or nil.

Scientific honesty seems to the best policy to convince people about the advantages of biotechnology for crop improvement. What to do? Scientists, farmers, consumers, and policy-makers should objectively assess the potential hazards of crop biotechnology in farming and food systems regarding the current situation and the likelihood that such hazards may occur.

For example scientists should explain to the people that gene recombination (or reassortment) already occurs in nature. However, the ecological success of viable recombinants after gene reassortment is unpredictable owing to the high fitness of current isolates. For this reason, more scientific research will be needed to identify unpredictable risks and the chances of their occurrence.

The need for profit, as in any other business, has attracted the interest of the private sector to defend their investments in crop biotechnology with patents, intellectual property rights, and new protection methods, *e.g.* 'terminator' technology that inhibits germination of self-pollinated seeds. This technology protection system prevents farmers from saving seeds from their harvest for further utilization as next season planting propagules.

Three genes, each with a specific promoter, are inserted into the 'terminator' plant. One of the genes (*e.g.* CRE/LOX system from bacteriophages) produces a recombinase that removes a spacer between the gene producing, for example, a ribosomal inhibitor protein and its promoter such as late embryonic abundance, which only becomes active during the late stages of embryo development. This spacer with specific recognition sites blocks the gene (for the ribosomal inhibitor protein) from being activated.

Another gene (*e.g.* tetracycline repressor system) produces a repressor that keeps off the recombinase gene until an outside stimulus is applied to the 'terminator' plant, *e.g.* a chemical such as the tetracycline, or temperature and osmotic shocks.

The United States Department of Agriculture (USDA) and a cotton seed enterprise jointly acquired a patent for this concept (U.S. patent 5,723,765). Two months after this patent was announced, one of the leading agro-chemical transnationals bought the cotton seed company, although one of its officers said that it may take many years before this 'terminator gene' idea becomes a proven technology in the seed industry.

Strategic alliances, joint ventures, research partnerships, new investments, company mergers, cross-ownerships, and take-overs in the seed and agro-chemical business have also been in the news in recent months. Likewise, some leading scientists are leaving their academic appointments to join the new private enterprises in plant biotechnology. These events are happening because the private sector wants to use biotechnology to accelerate its growth in agri-business in the short-term. Nonetheless, funds to support basic and strategic research by public researchers are needed for a long-term sustainable transfer of public goods (both knowledge and technology) to the private sector or other users.

BIOINFORMATICS

Another important factor in the successes of the genetic improvement of crops was the development of fast and more reliable computers, which allowed easier management and analysis of data as well as publication of scientific reports. The impact of the informatic revolution in crop improvement can be partially assessed by counting the number of publications indexed in Plant Breeding Abstracts (CAB International, Wallingford, Oxon, UK).

There was ca. 22-fold increase of publications in the 1930-1997 period. It was in the 1970s that indexed publications in plant breeding exceeded 10,000 per year. More publications and easy means for retrieving this information accounted for such growth of knowledge dissemination in plant genetics and breeding. Today, rapid information exchange has been facilitated with electronic mail and access to the internet to read electronic publications such as this journal. Nowadays, information technology and DNA science are beginning to fuse into a single operation.

Computers are deciphering, and organizing the huge genetic information that may become "the raw resource of the emerging biotech economy" in the next century. Scientists working in the new field of "bioinformatics" are developing biological data banks to download the genetic information accumulated during millions of years of life evolution, and perhaps reconstruct some of the living organisms of the natural world.

Plant Genomics

This new term, defined by the development of biotechnology, refers to the investigations of whole genomes by integrating genetics with informatics and automated systems. Genomic research aims to elucidate the structure, function and evolution of past and present genomes. Some of the most dynamic fields concerning agriculture are the sequencing of plant genomes, comparative mapping across species with genetic markers, and objective assisted breeding after identifying candidate genes or chromosome regions for further manipulations. As a result of genomics, the concept of gene pools has been enlarged to include transgenes and native exotic gene pools that are becoming available through comparative analysis of plant biological repertoires. Understanding the biological traits of one species may enhance the ability to achieve high productivity or better product quality in another organism. DNA markers and gene sequencing provides quantitative means to determine the extent of genetic diversity and to establish objective phylogenetic relationships among organisms. 'Gene chips' and transposon tagging will provide new dimensions for investigating gene expression. Molecular biologists will study not only individual genes but how circuits of interacting genes in different pathways control the spectrum of genetic diversity in any crop species. For example, more information will be available on why plant resistance genes are clustered together, or what candidate genes should be considered when manipulating quantitative trait loci (QTL) for crop improvement.

FARMING IN ENVIRONMENTALLY FRIENDLY SYSTEMS

The aims of applied plant science research for agriculture are to enhance crop yields, improve food quality, and preserve the environment where human beings and other organisms live. The best way for conservation of plant biodiversity and its environment, would be to achieve high crop productivity per unit area. In this regard, Briggs (1998) reported that as yields treble, soil erosion per ton of food decreases by two-thirds. There has been a significant yield improvement owing to enhanced crop husbandry, but in the next years progress will be achieved by changing plants that could be more suitable to sustainable and environmentally friendly farming systems. Agro-chemical corporations are developing pest and disease resistant transgenic crops to avoid pollution with pesticides in the farming system. Furthermore, food quality will become more important than crop productivity in a wealthy society. Consumers will prefer transgenic crops if they have the desired characteristics.

In the next decades meiotic-based breeding will still generate cultivars for farmers. Genetic improvement through biotechnology needs conventional breeding because (1) the elite cultivars will be the parents of the next generation of improved genotypes, (2) field testing across locations or cropping systems and over years will be needed to determine the best selections due to the genotype-by-environment interaction. As stated by Briggs (1998), "transgenes must be viewed as improvements rather than replacements for elite germplasm". Indeed, genetic engineering may provide a means to add value by introducing synthetic or natural genes that enhance crop quality and yield, as well as protect the plant against pest and diseases. Farmers will pay more for transgenic crop propagules if they obtain extra-income after adopting biotech-derived products. For example, seeds of insect resistant transgenic crops will be more expensive than those of available cultivars but the farmer will not need to apply pesticides in their transgenic fields. Of course, patents make transgenic seeds more expensive but also farmer's benefits may be higher.

GENE BANKS, DNA BANKING AND VIRTUAL PLANT BREEDING

The sequencing of crop genomes opened new frontiers in conservation of plant biodiversity and its genetic enhancement. The advances in gene isolation and sequencing in many plant species allows to envisage that within a few years, gene-bank curators may replace their large cold stores of seeds with crop DNA sequences that will be electronically stored.

The characterization of plant genomes will ultimately create a true gene bank, which should possess a large and accessible gene inventory of today's non-characterized crop gene pools. Of course, seed banks of comprehensively investigated stocks should remain because geneticists and plant breeders, the main users of gene banks, will need this germplasm for their work. Genomics may accelerate the utilization of candidate genes available at these gene banks through transformation without barriers across plant species or other living kingdoms. Nonetheless, genetic engineering should be seen as one of the methods of plant breeding that permits the direct alteration and re-building of a crop population. "Shutting-off" genes coding for undesired characteristics may be another application of transgenics in crop improvement.

Plant breeders will change their modus operandi with the development of objective marker-assisted introgression and selection methods. Backcross breeding will be shortened by eliminating undesired chromosome segments (also known as linkage drags) of the donor parent or selecting for more chromosome regions of the recurrent parent. Parents of elite crosses may be chosen based on a combination of DNA markers and phenotypic assessment in a selection index, such as best linear unbiased predictors. To achieve success in these endeavours, cheap, easy, decentralized, and rapid diagnostic marker procedures are required.

There are many areas of basic and strategic research in plant breeding and genetics that are being facilitated by marker-aided analysis. With molecular markers, plant biologists are reviewing crop evolution and gathering new knowledge. Such information should be incorporated into genetic enhancement programmes, especially those with an evolutionary breeding scheme. Likewise, plant ideotypes for each crop should drive the work of plant breeders. Specific plant morphotypes have been defined in rice and wheat based on accumulated knowledge of crop physiology and crop protection. The needed characteristics required to develop improved plant prototypes ensuing from such a 'virtual breeding' approach may be available in gene banks of the crop or in those of other species. Otherwise, breeders may obtain novel transgenes to develop the required ideotype.

Nowadays, the finding of new genes that add value to agricultural products seems to be very important in the private agri-business. Unique gene databases are being assembled by the industry with the massive amount of data generated by genomics research. A new term 'biosource' was coined recently to refer to a fast and effective licensed technology of pinpointing genes. With this method, a 'benign' virus infects a plant with a specific gene that allows researchers to observe directly its phenotype. Biosource replaces the standard time-consuming approach of first mapping a gene to subsequently determine its exact function. Gene identification in DNA libraries coupled with biosource technology and an enhanced ability to put genes into plants will be routine for improving crops in the next decade.

Genomics may provide a means for the elucidation of important functions that are essential for crop adaptedness. Regions of the world should be mapped by combining data of geographical information systems, crop performance, and genome characterization in each environment. In this way, plant breeders can develop new cultivars with the appropriate genes that improve fitness of the promising selections. Fine-tuning plant responses to distinct environments may enhance crop productivity. Development of cultivars with a wide range of adaptation will allow farming in marginal lands.

Likewise, research advances in gene regulation, especially those processses concerning plant development patterns, will help breeders to fit genotypes in specific environments. Photoperiod insensitivity, flowering initiation, vernalization, cold acclimation, heat tolerance, host response to parasites and predators, are some of the characteristics in which advanced knowledge may be acquired by combining molecular biology, plant physiology and anatomy, crop protection, and genomics. Multidisciplinary cooperation among researchers will provide the required holistic approach to facilitate research progress in these subjects.

Pharming and Farmer-ceuticals

Growth of cities in the developed world has already replaced farmland with shopping malls, parking lots, and housing developments. Peri-urban agriculture and home gardening are also becoming very important for national food security in the developing world as a result of rapid urban expansion. Hence, new cultivars will be needed to fit into intensive production systems, which may provide the food required to satisfy urban world demands of the next century. Specific plant architecture, tolerance to urban pollution, efficient nutrient uptake, and crop acclimatization to new substrates for growing are, among others, the plant characteristics required for this kind of agriculture. Genes controlling these characteristics may be available in gene banks for further cross breeding, which can be assisted by genomics. Peri-urban and home garden "farmers" will have to adapt to new demands from emerging urban populations with higher income. These consumers may request a more varied diet. For example, food crops with low fats, and high in specific amino acids may be needed to satisfy people who wish to change their eating habits. If genes controlling these characteristics do not exist in a specific crop pool they may be incorporated into the breeding pool using transgenics.

Some publications anticipated that in the next millennium food will not need to be harvested from farmer's fields. Tissue culture of certain parts of the plant may provide a means to achieve success in this endeavour. For example, edible portions of fruit crops could be grown in vitro. A steady and cheap supply of these edible plant parts will be required in this new agri-business. It will take some time before such a process can be scaled-up for commercial output. Nonetheless, a patent was submitted in 1991 by a Californian biotech company for producing a vanilla extract through cell culture. Of course, this technique will not replace farming as we know it today. This biotechnique, as well as other new farming methods, offers a means for new ways of producing food, feed or fibre.

Often plants provide the raw materials for agro-industry, and not only for food or fibre processing. Active ingredients of plants have been transformed into commercial products such as medicines, solvents, dyes, and non-cooking oils for many years. Hence, it would not be surprising to see, in few years from now, entire farms without food crops but growing transgenic plants to produce new products, *e.g.* edible plastic from peas or plant oils to manufacture hydraulic fluids and nylon. This new rural activity may result in important changes in the national economic sector.

'Pharming' has been added to the dictionary to indicate a new kind of system to obtain medicines. For example, oral vaccines appear to be a convenient delivery system for vaccination throughout the world. Biotechnology has been used to engineer plants that contain a gene derived from a human pathogen. An antigenic protein encoded by this foreign DNA can accumulate in the resultant plant tissues. Results from pre-clinical trials showed that antigenic proteins harvested from transgenic plants were able to keep the immunogenic properties if purified. These antigenic proteins caused the production of specific antibodies in injected mice. Mice, which ate these transgenic plant tissues, also showed also a mucosal immune response. Arakawa *et al.* (1998) recently demonstrated the ability of transgenic food crops to induce protective immunity in mice against a bacterial enterotoxin such as cholera toxin B subunit pentamer with affinity for GMI-ganglioside. Also, potato tubers have been used successfully as a biofactory for high-level output of a recombinant single chain antibody.

PLANT BREEDING FOR A WORLD COMMUNITY

There is enormous potential for plant breeding to benefit less developed parts of the world, where food security is critical as populations continue to increase.

In particular, advances in genetic technology may offer more versatile solutions than conventional systems of crop improvement. For example, scientists in Britain are pioneering important research to develop salt and drought tolerance in crop plants. This may help to alleviate the devastating effects of crop failure in arid regions such as sub-Saharan Africa. Private and public sector initiatives to share knowledge and expertise are already in place. They include commitments to provide Vitamin A enhanced rice and virus resistance in sweet potatoes free of charge to developing countries.

International arrangements to recognize genetic resources, as the sovereign property of individual nation states will stimulate further programmes of technology transfer and benefit sharing between industrialized and less developed regions of the world.

EFFECTS OF MODERN PLANT BREEDING ON ENVIRONMENT

Some environmentalists are concerned that genes from genetically modified crops could escape and transfer to other species with unwanted consequences. For example, it is argued that herbicide-resistant crops could cross with weedy relatives to create a new strain of 'super weed'. In practice, since the reproductive systems of genetically modified and conventionally bred crops are identical, the behaviour of domesticated crop plants is unlikely to be affected by single gene changes.

In ten years of worldwide field trials there have been no adverse reports of genetically modified crops spreading in the environment. Furthermore, resistance to weed killers has been available for decades in a number of conventionally bred varieties without causing any such problem. The development conventionally bred crops-nevertheless the technology is subject to much tighter controls. In the long history of plant breeding, the strict regulations applied to the development and use of genetically modified crops is unprecedented. In both the laboratory and the field, each genetically modified crop must go through a rigorous process of monitoring and evaluation before it can reach the customer. Extensive and ongoing evaluation of genetically modified crops has shown that the technology presents no new food safety risks. Indeed, biotechnology will enable breeders to develop food crops with improved nutritional value and better keeping qualities. Enhanced disease and pest resistance will also reduce pesticide residues.

AIM OF PLANT BREEDING

The constant aim of plant breeding is to improve the quality, diversity and performance of agricultural and horticultural crops. The overriding objective is to develop plants better adapted to human needs. The evolution of plant breeding is a classic example of how improved biological understanding has been adapted to provide more effective methods of meeting the demands of a changing world. Plant breeding has existed in its most primitive form since the first farmers saved the seeds of their best plants from one season to the next more than 10,000 years ago. Over the centuries, this selection process has gradually become more scientific, bringing major improvements in the yield, quality and diversity of crops grown worldwide.

In the 19th century Gregor Mendel established the basic principles of plant genetics. Gregor Mendel in the 19th century laid the foundations for traditional plant breeding, in which selected parents are cross-pollinated to combine certain desired characteristics, such as high yield and resistance to pests and disease. He discovered that units of material, which are transferred from one generation to the next, determine inherited traits.

The plant breeder's aim is to reassemble these units of inheritance, known as genes, to produce crops with improved characteristics. In practice, this is a complex and time-consuming process. Each plant contains many thousands of genes, and the plant breeder is seeking to combine a range of desirable traits in one plant to produce a successful variety. Conventional breeding involves crossing selected parent plants, chosen because they have desirable characteristics such as high yield or disease resistance. The breeder's skill lies in selecting the best plants from the many and varied offspring. These are grown on and tested in subsequent years. Typically this involves examining thousands of individual plants for different characteristics ranging from agronomic performance to end-use quality.

Developing a new variety can take up to 15 years for wheat, 18 years for potatoes, even longer for some crops. The scope of conventional plant breeding has increased with improvements in technology. In the laboratory, chemical and mechanical techniques are used to speed up the selection process and remove natural barriers to cross-fertilization, for example between different crop species.

All forms of plant breeding involve the improvement of crops by selection. Since man began farming he has selected seeds from the best plants for the next generation. Over the centuries, this selection process has become more scientific, bringing major improvements in the yield, quality and diversity of agricultural and horticultural crops. Genes-units of hereditary material that are transferred from one generation to the next, determine these characteristics. Since the plant contains many thousands of genes, and the breeder is seeking to combine a range of traits in one plant, developing a successful variety can be an extremely lengthy process-up to 12 years in the case of cereals.

Using biological knowledge derived from cross-pollination, plant breeders have developed physical and chemically assisted ways of enhancing the speed, accuracy and scope of the selection process. Breeding systems involving grafting and hybridization, for example, have extended breeders' capabilities at the whole crop level, while more recent cell-based techniques enable breeders to operate at the level of individual cells and their chromosomes

A major objective in modern plant breeding is the making of crop varieties with the highest possible yield potential. Yield potential is defined as "the yield of a crop when growth is not limited by water or nutrients, pests, diseases, or weeds". For farmers, whose crops are indeed limited by such constraints, this yield concept may not be seen as the most relevant objective. Relieving crops of all sorts of environmental stress, leaves light and temperature, together with varietal characteristics, as the only determinants of yield. That achieved, the same high yielding variety can be used all over a climatic zone, and target area for the variety is enormous. If stress can not be eliminated, however, adaptation to a usually site-specific environment becomes necessary. In that case, the target area for each variety will be small.

Do these different perspectives warrant different approaches to plant breeding? Conventional wisdom says no. In experiments where varieties are tested at various levels of inputs, the same high yielding variety usually comes out as top yielder at all levels. Therefore, plant breeders often claim that their varieties not only perform excellently under high-input conditions, but would also be better than traditional varieties in a low-input environment. However, when these trials are taken outside of experimental farms and the varieties are tested under local or farm conditions, researchers discover what is called 'crossover in performance'. At a certain level of stress there is a crossover point beyond which local varieties or landraces perform better than the high yielding varieties Farmers who experience such situations are not a tiny minority.

To some degree, most if not all farmers have to cope with local stress conditions. The stress-free environment is hard to achieve, and even harder to sustain. In many favourable areas, farmers abandon the high-input technology for economic reasons or because of ecological problems, thus increasing the need for locally-adapted germplasm. But modern plant breeding, whether public or private, can not supply adapted germplasm everywhere. Only a system of local seed selection can ensure that. And that means devolution of plant breeding.

Can such decentralized breeding be compatible with development needs in a changing world and meet economic aspirations in a poor society? If the answer to these questions is going to be 'yes', the decentralized breeding must be able to take advantage of the power of science as well as of the capacities of local communities. Three aspects should be considered and made mutually compatible: breeding technology, participatory research methods, and organization at community level.

HISTORY

Plant breeding is the purposeful manipulation of plant species in order to create desired genotypes and phenotypes for specific purposes. This manipulation involves either controlled pollination, genetic engineering, or both, followed by artificial selection of progeny. *Plant breeding* often, but not always, leads to plant domestication.

Plant breeding has been practiced for thousands of years, since near the beginning of human civilization. It is now practiced worldwide by government institutions and commercial enterprises. International development agencies believe that breeding new crops is important for ensuring food security and developing practices of sustainable agriculture through the development of crops suitable for their environment.

Domestication

Domestication of plants is an artificial selection process conducted by humans to produce plants that have fewer undesireable traits of wild plants, and which renders them dependent on artificial (usually enhanced) environments for their continued existence. The practice is estimated to date back 9,000-11,000 years. Many crops in present day cultivation are the result of domestication in ancient times, about 5,000 years ago in the Old World and 3,000 years ago in the New World. In the Neolithic period, domestication took a minimum of 1,000 years and a maximum of 7,000 years. Today, all of our principal food crops come from domesticated varieties.

A cultivated crop species that has evolved from wild populations due to selective pressures from traditional farmers is called a landrace. Landraces, which can be the result of natural forces or domestication, are plants (or animals) that are ideally suited to a particular region or environment. An example are the landraces of rice, *Oryza sativa* subspecies *indica,* which was developed in South Asia, and *Oryza sativa* subspecies *japonica,* which was developed in China.

Classical plant breeding uses deliberate interbreeding (crossing) of closely or distantly related species to produce new crops with desirable properties. Plants are crossed to introduce traits/genes from one species into a new genetic background. For example, a mildew resistant pea may be crossed with a high-yielding but susceptible pea, the goal of the cross being to introduce mildew resistance without losing the high-yield characteristics. Progeny from the cross would then be crossed with the high-yielding parent to ensure that the progeny were most like the high-yielding parent, (backcrossing), the progeny from that cross would be tested for yield and mildew resistance and high-yielding resistant plants would be further developed. Plants may also be crossed with themselves to produce inbred varieties for breeding.

Classical breeding relies on homologous recombination of two genomes to generate genetic diversity. The Classical plant breeder may also makes use of a number of *in vitro* techniques such as protoplas fusion, embryo rescue or mutagenisis to generate diversity and produce plants that would not exist in nature.

Traits that breeders' have tried to incorporate into crop plants in the last 100 years include:

- Increased quality and yield of the crop
- Increased tolerance of environmental pressures (salinity, extreme temperature, drought)
- Resistance to viruses, fungi and bacteria
- Increased tolerance to insect pests
- Increased tolerance of herbicides

Before World War II

Intraspecific hybridization within a plant species was demonstrated by Charles Darwin and Gregor Mendel, and was further developed by geneticists and plant breeders. In the early 20th century, plant breeders realised that Mendel's findings on the non-random nature of inheritance could be applied to seedling populations produced through deliberate pollinations to predict the frequencies of different types.

In 1908, George Harrison Shull described heterosis, also known as hybrid vigour. Heterosis describes the tendency of the progeny of a specific cross to outperform both parents. The detection of the usefulness of heterosis for plant breeding has lead to the development of inbred lines that reveal a heterotic yield advantage when they are crossed. Maize was the first species where heterosis was widely used to produce hybrids.

By the 1920s, statistical methods were developed to analyse gene action and distinguish heritable variation from variation caused by environment. In 1933, another important breeding technique, cytoplasmic male sterility (CMS), developed in maize, was described by Marcus Morton Rhoades. CMS is a maternally inherited trait that makes the plant produce sterile pollen, enabling the production of hybrids and removing the need for detasseling maize plants.

These early breeding techniques resulted in large yield increase in the United States in the early 20th century. Similar yield increases were not produced elsewhere until after World War II, the Green Revolution increased crop production in the developing world in the 1960s.

After World War II

Following World War II a number of techniques were developed that allowed plant breeders to hybridize distantly related species, and artificially induce genetic diversity.

When distantly related species are crossed, plant breeders make use of a number of plant tissue culture techniques to produce progeny from other wise fruitless mating. Interspecific and intergeneric hybrids are produced from a cross of related species or genera that do not normally sexually reproduce with each other. These crosses are referred to as *Wide crosses*. The cereal triticale is a wheat and rye hybrid. The first generation created from the cross was sterile, so the cell division inhibitor colchicine was used to double the number of chromosomes in the cell. Cells with an uneven number of chromosomes are sterile.

Failure to produce a hybrid may be due to pre- or post-fertilization incompatibility. If fertilization is possible between two species or genera, the hybrid embryo may abort before maturation. If this does occur the embryo resulting from an interspecific or intergeneric cross can sometimes be rescued and cultured to produce a whole plant. Such a method is referred to as *Embryo Rescue*. This technique has been used to produce new rice for Africa, an interspecific cross of Asian rice *(Oryza sativa)* and African rice *(Oryza glaberrima)*. Hybrids may also be produced by a technique called protoplast fusion. In this case protoplasts are fused, usually in an electric field. Viable recombinants can be regenerated in culture.

Chemical mutagens like EMS and DMSO, radiation and transposons are used to generate mutants with desirable traits to be bred with other cultivars. Classical plant breeders also generate genetic diversity within a species by exploiting a process called somaclonal variation, which occurs in plants produced from tissue culture, particularly plants derived from callus. Induced polyploidy, and the addition or removal of chromosomes using a technique called chromosome engineering may also be used.

When a desirable trait has been bred into a species, a number of crosses to the favoured parent are made to make the new plant as similar as the parent as possible. Returning to the example of the mildew resistant pea being crossed with a high-yielding but susceptible pea, to make the mildew resistant progeny of the cross most like the high-yielding parent, the progeny will be crossed back to that parent for several generations. This process removes most of the genetic contribution of the mildew resistant parent. Classical breeding is therefore a cyclical process.

It should be noted that with classical breeding techniques, the breeder does not know exactly what genes have been introduced to the new cultivars. Some scientists therefore argue that plants produced by classical breeding methods should undergo the same safety testing regime as genetically modified plants. There have been instances where plants bred using classical techniques have been unsuitable for human consumption, for example the poison solanine was accidentally re-introduced into varieties of potato though plant breeding.

Exploiting heterogeneity and crop evolution in farmers' fields are outside the scope of most plant breeding research. One exceptional experiment, however, has shed some scientific light on the issue. It was started at the University of California (UC) in 1928. Composite cross populations of barley were produced, some of which were extremely diverse in origin of sources. These populations were exposed to continuous natural selection in current modern farming environments and became the subject of studies during the career span of several generations of UC professors.

It appears that after low yields in initial years, the composite cross populations gradually improved in performance and eventually became quite good yielders, with excellent yield stability and disease resistance. These results inspired Suneson to propose an evolutionary plant breeding method. After assessment of later generations of the same material, Soliman and Allard concluded that such evolutionary breeding "is unwarranted" if yield potential is the major goal. However, if disease resistance and yield stability are two major objectives, "the composite cross approach is an efficient method". This amounts to saying that a major part of world agriculture, many high-input systems included, could be well served by this approach.

In short, this means constructing a body of broadly diversified germplasm and exposing it to natural selection in areas of contemplated use. For those who are familiar with traditional farmers' breeding, this may sound like reinventing the wheel. In fact it is an improvement of the old wheel of plant breeding.

The first step, constructing a body of of broadly diversified germplasm, is not all that straightforward. Science has access to world collections and information sources that are unavailable to farmers. A research institute can chose relevant germplasm and make composite cross populations with an evolutionary potential, most probably far beyond that of locally available varieties.

The immediate outcome, the early generation composite cross population, will be unadapted everywhere and is likely to yield poorly. With time, however, recombinations and natural sorting will improve the adaptation, and, according to the Californian experience, narrow the gap with commercial varieties. The long term outcome could be populations that outperform commercial varieties in disease resistance and yield stability and that may be used as a source of artificial selection for high yield.

The disease resistance appears to have evolved through the building up of polygenic complexes. Therefore, it provides a durable resistance as opposed to the monogenic and, therefore, mostly non-durable resistance usually bred into commercial pure line varieties. The stability, at least to some degree, depends on the buffering effect of crop heterogeneity. The Californian experiment shows that when the population is propagated in isolation for a very long time, diversity will start declining, resulting eventually in reduced stability.

These experimental findings into the context of current development needs, a few conclusions can be drawn. We need breeding populations with a very high evolutionary potential, and these populations must be exposed to the stress conditions of, or similar to, current farm environments. Furthermore, a certain level of diversity within populations must be maintained in order to sustain evolutionary potential and yield stability.

An Age-old Tradition

Modern plant breeding is a sophisticated, high investment business, but its origins stretch back thousands of years to primitive farmers who selected the best plants in one year to provide seed for their next crop. This selective breeding was the first human refinement of natural plant evolution. Recent scientific and technological developments have allowed a greater rate of improvement. It was Gregor Mendel who, in the mid- 19th century, first provided a scientific explanation of genetic inheritance. His conclusions on the relationship between inherited characteristics in the offspring and the genetic makeup of the parents were the theoretical basis for classical plant breeding. Mendel's work went largely unrecognized in his own lifetime, and it was not until the early 20th century that it was rediscovered to become the basis of modern scientific plant breeding.

A Modern Industry

Until the early 1960s, plant breeding in Britain was largely confined to publicly funded research. This situation changed dramatically in the mid-

1960s, with the passing into UK law of the 1964 Plant Varieties and Seeds Act. This legislation introduced a system of royalty payments on individual plant varieties, known as Plant Breeders' Rights, and triggered a rapid expansion of plant breeding as a commercial enterprise in its own right. Today, much of the basic research into crop science is still conducted by public sector research organisations, but the majority of commercial plant breeding takes place within the private sector. Some

60 plant breeding companies, based in the UK, are active across the entire spectrum of plant species from the major arable crops through to ornamental garden shrubs and flowers. In total, the plant breeding sector employs around 5,000 people, and supports a further 5,000 jobs in seed production and distribution. Plant breeding remains a vital industry to keep Britain competitive on world markets. The need for new varieties, adapted to our unique growing conditions, is never ending, driven by the challenges of new disease pressures, changing market requirements and shifts in agricultural and environmental policies.

- *Maize*: Maize crops used for grain and animal fodder are derived from wild races originating in Central America
- *Sugar beet*: Modern varieties of sugar beet have been developed from wild ancestors native to Central Europe.
- *Potatoes*: Wild ancestors of the modern potato still grow in parts of South America.
- *Wheat and Barley*: Small grain cereals, the mainstay of UK crop production, are derived from wild grasses of the Middle East.
- *Oilseed Rape*: Like many crops within the brassica family, oilseed rape has its origins in wild species native to China.

Participatory Plant Breeding

Certain trends have made the world ripe for adoption of participatory plant breeding methods.

- The shielding of crops from environmental stress in high-input systems is facing increasing economic and ecological problems. Scientists are changing their attitudes, and a new paradigm is being formulated. Instead of modifying the environment to suit the requirement of high yielding varieties, the varieties need to be modified to suit the environment.
- The claim that modern varieties can be made broadly adapted and be superior across most farming environments within an ecogeographic region is being challenged.
- If relevant diversity exists in a locale, the combined action of natural and artificial selection within a local environment may be an efficient breeding method. Experiments show that this may work also in a fertilizer-intensive system.

- In recent years farmer groups working with local seeds have been organized all over the world. They are not primarily conservers of old seeds. They want their seeds to be improved, in an evolutionary, slow and steady way, and under their own control.
- Finally, participatory methods have been developed in order to facilitate the involvement of farmers together with scientists as active and equal partners in research to generate relevant farm technology. Such methods can be applied also to plant breeding.

Commercial seeds often diffuse into areas where the traditional seed supply system is still predominant. Farmers try them with an open mind and adopt or reject them according to their own criteria. If grown and multiplied in the villages, diversity will start to appear within them and local reselection will be possible. In that way commercial varieties eventually might become like landraces. It is also commonly observed that farmers change and exchange seeds. A traditional farming system rarely functions as an environment for the static preservation of old landraces.

Often practices range from neglect to very simple mass selection, but with a few scattered individuals who devote an exceptional amount of effort to the maintenance or improvement of seed quality. These exceptional persons, very often women, may be the source of good seeds for others in the community. Once such a community is organized for seed management and improvement, it becomes possible for it to establish links to scientific institutions.

Community resources for participatory approaches to seed management and breeding are not limited to indigenous culture and traditional practices. The educational status and experiences of modern farmers may also be turned into a resource for community action. In the Philippine group, a number of the members were high school graduates and a few had a university degree. And moreover, most of them had a couple of decades' experience with modern input-intensive farming. Seed activities opened their minds towards the traditional societies, towards themselves and towards the modern world. The traditional societies supplied them with their seeds, and through the seeds, they learned to appreciate the values and achievements of these societies. They discovered their own potential, and saw that the outside world could bring more than technology packets: it could bring knowledge and ideas to be exploited and further developed by themselves.

CONVENTIONAL BREEDING

Conventional plant breeding involves crossing carefully chosen parent plants, then selecting the best plants from the resulting offspring to be grown on for further selection. For cereals, hundreds of individual crosses are carried out by hand to create seed for the first filial (or F1) generation. The resulting

F1 plants are uniform, but in the following generation several hundred thousand different plants are produced. Because of the way genes work, the new combinations produced from each cross are not revealed until the second (F2) generation.

It is this enormous diversity of new gene combinations that may hold the key to a successful new variety. The plant breeder's task is to select the plants most likely to meet his breeding objectives. Seed from the best of these F2 plants is grown on in small rows or plots and the best plants again selected–this process is repeated year after year until only the very best plants remain. As promising new lines emerge, tests are conducted on each plot to assess factors such as yield, disease resistance and endues quality.

Once the best lines are purified to ensure that every plant has the same characteristics, the process of multiplying seed begins. These 'inbred' lines are then ready for entry into official trials, some six to ten years after the initial cross.

HYBRID BREEDING

For some crop species, the seed supplied to growers is that produced from the first cross between selected parents. The resulting varieties, known as F1 hybrids, offer potential advantages in crop performance. Hybrid breeding is widely used to produce varieties of field vegetables, maize and oilseed rape. F1 hybrids are unique in expressing 'hybrid vigour' in the growing crops for a single year. This may result in higher yields, greater uniformity, or improvements in quality. Unlike inbred lines, F1 hybrids do not breed true year after year and their performance gains are not maintained in subsequent generations.

IMPROVED BREEDING

With increasing knowledge and improved technology, breeders have developed ways to enhance the speed, accuracy and scope of the breeding process.

There are several ways to reduce the lengthy interval between the first cross of selected parents and establishing true breeding lines of promising new varieties. For example, maintaining parallel selection programmes in northern and southern hemispheres allows two generations to be produced each year. Single seed descent produces very small plants under restricted growth conditions. Large numbers of these plants are cultivated in artificial growth rooms, with two or more generations produced in a year. In potatoes, mini-tuber breeding speeds up the slow multiplication process by producing miniature plants under greenhouse conditions.

More recent laboratory techniques enable breeders to operate at the level of individual cells and their chromosomes. In certain crop species, such as potatoes and oilseed rape, it is possible to produce new varieties in the laboratory through

Protoplast Fusion

Individual plant cells with their outer walls removed are fused, and the fused cells then induced to divide and grow in a culture medium. Whole plants are eventually regenerated containing new combinations of genes from the two parents.

Embryo Rescue and Assisted Pollination

Allow breeders to expand the range of available characters by making crosses between species, which would not produce viable offspring outside the laboratory.

Double Haploid Breeding

Enables breeders to produce genetically uniform lines within one generation. This effectively by-passes the lengthy process of self pollination and selection normally required to produce true breeding plants. Latest developments in genetic science have greatly improved our understanding of how plants behave, offering additional ways to enhance the breeding process.

Genomics

By mapping the genetic makeup, or genome, of crop species, scientists can identify the exact position and function of individual genes. Genome mapping has revealed striking similarities in the genomes of different crop species, such as rice, wheat, barley and rye. This information is already helping to broaden the scope and precision of current breeding programmes.

Marker Assisted Breeding

Allows breeders to determine whether desired traits are present in a new variety at an early stage in the breeding programme.

GENETIC MODIFICATION

Genetic modification of plants is achieved by adding a specific gene or genes to a plant, or by knocking out a gene with RNAi, to produce a desirable phenotype. The resulting plants are often referred to as transgenic plants. Genetic modification can produce a plant with the desired trait or traits faster than classical breeding because the majority of the plant's genome is not altered.

To genetically modify a plant, a genetic construct must be designed so that the gene to be added or knocked-out will be expressed by the plant. To do this, a promoter to drive transcription and a termination sequence to stop transcription of the new gene, and the gene of genes of interest must be introduced to the plant. A marker for the selection of transformed plants is also included. In the laboratory, antibiotic resistance is a commonly used marker: plants that have been successfully transformed will grow on media containing antibiotics; plants that have not been transformed will die. In some instances markers for selection are removed by backcrossing with the parent plant prior to commercial release.

The construct can be inserted in the plant genome by genetic recombination using the bacteria Agrobacterium tumefaciens or *A. rhizogenes*, or by direct methods like the gene gun or microinjection. Using plant viruses to insert genetic constructs into plants is also a possibility, but the technique is limited by the host range of the virus. For example, Cauliflower mosaic virus (CaMV) only infects cauliflower and related species. Another limitation of viral vectors is that the virus is not usually passed on the progeny, so every plant has to be inoculated.

The majority of commercially released transgenic plants, are currently limited to plants that have introduced resistance to insect pests and herbicides. Insect resistance is achieved through incorporation of a gene from Bacillus thuringiensis (Bt) that encodes a protein that is toxic to some insects. For example, if cotton pest the cotton bollworm feeds on Bt cotton it will ingest the toxin and die. Herbicides usually work by binding to certain plant enzymes and inhibiting their action. The enzymes that the herbicide inhibits are known as the herbicides *target site*. Herbicide resistance can be engineered into crops by expressing a version of *target site* protein that is not inhibited by the herbicide. This is the method used to produce glyphosate resistant crop plants.Genetic modification of plants that can produce pharmaceuticals (and industrial chemicals), sometimes called *pharmacrops*, is a rather radical new area of plant breeding

Proteomics

Allows breeders to understand how genes behave in different parts of the plant and under different growing conditions.

Maintaining Genetic Diversity

Maintaining biodiversity is central to the process of crop improvement. It is in every breeder's interest to ensure that the gene pool from which new traits are selected remains as extensive as possible. Plant breeders created the first gene banks in the 1930s to conserve the valuable genetic diversity within past and present varieties, as well as landraces and wild relatives of cultivated crop species. Plant breeding is integral to ongoing initiatives to identify, classify and conserve existing biodiversity.

Testing Plant Varieties

Before any new crop variety can be placed on the market, it must undergo statutory testing under a process known as National Listing. Successful varieties are placed on a National List or register of varieties approved for marketing. National Listing rules are determined on a European basis, and apply to all the major agricultural and vegetable crop species. Official trials are conducted, in most cases for a minimum of two years, to test each new variety for a range of characteristics which together determine its uniqueness, its genetic uniformity, and its value to growers and the rest of the food chain. National Listing is extremely rigorous–the majority of varieties entered do not complete the process. In winter wheat, for example, only a quarter of varieties entered for National Listing during the 1990s were finally approved.

National Listing–DUS

All varieties entered for National Listing are assessed for Distinctness, Uniformity and Stability (DUS). In the case of cereals, some 30 individual characteristics of the plant are inspected to verify that it is distinct, ie clearly distinguishable from other varieties, that its characteristics are uniform from one plant to another, and that the variety is stable in that it breeds true to type from one generation to the next.

National Listing–VCU

For agricultural crops, National Listing also involves trials to establish a variety's Value for Cultivation and Use (VCU). This provides an assurance to growers that only varieties with improved performance or end-use quality can be officially approved for sale.

Recommended and Descriptive Lists

Once a new variety has been added to the National List, it is cleared for marketing. However, its success in the market place is by no means guaranteed. Further independent trials are conducted each year to compare the performance and quality of the best varieties. These trials provide the basis for detailed information and advice to growers and their customers.

As shown in the following illustration for winter wheat, the process of statutory and commercial evaluation can take up to five years. Only a handful of varieties clear all these hurdles, which are in addition to the many years spent testing and selecting in the breeder's own trials programme.

Genetically Modified Varieties

No GM crops can be marketed until they have been assessed and approved in terms of human health, food safety and the environment. The cross-border movement of genetically modified organisms (GMOs) is regulated at a global level under the internationally agreed Biosafety Protocol.

Variety Maintenance and Identity Preservation

As well as developing new varieties, plant breeders maintain the genetic purity of existing lines and pre-commercial seed supplies year by year. This process is costly and time-consuming, but essential to maintain the quality and performance of each variety.

For cereals, variety maintenance begins after a few years of selection trials, when the promise of a variety is just emerging. At that stage, all that exists of what may become a widely grown variety is a single row containing around 100 plants. The breeder then bulks up supplies of the purified lines of breeder's seed into prebasic and then basic seed. Each year specialist seed growers are used to grow basic seed for the first generation of certified seed–or C1 seed. After one more year this becomes C2 seed, the main source of certified seed used by farmers.

The plant breeder continuously maintains breeder's seed for the process of multiplication through pre-basic, basic, C1 and C2 seed to ensure the variety's performance and quality year after year. Greater emphasis is now being placed on preserving the identity of individual varieties after harvest, both to conserve quality characteristics and to meet consumer demands for assurances about the integrity and traceability of their food.

Seed Certification

Seed of an approved variety can only be marketed if it meets strict quality criteria Seed quality standards are laid down in UK and EU law, and policed by agencies appointed by Government. The UK's official seed certification system offers an independent assurance of quality to growers. Minimum standards apply for varietal identity, purity and germination capacity. In addition, strict limits apply to seed-borne diseases, and the presence of physical impurities such as weed seeds.

Around 9% of the UK arable area is used to multiply the pure lines of seed from the plant breeder into certified seed. Several thousand individual crops are involved; each grown under specific management regimes to ensure the purity and integrity of the resulting seed is maintained. To gain certification, every seed crop must undergo crop inspection and seed testing. Seed certification underpins the health and purity status of the major arable crops in Britain. It offers an independent benchmark of quality on which buyers of seed and their customers depend.

Farm-saved Seed

For certain crop species–particularly small-grain cereals–growers can opt to save their own seed for sowing the following year provided care is taken to ensure that the crop remains healthy and free from impurities, and that the resulting seed is carefully conditioned and cleaned. Without independent testing for germination and freedom from seed borne diseases, however, farm-saved seed can harbors risks to growers that may not become apparent until well into the growing season. Many growers choose certified seed for the peace of mind that quality and performance is independently assured.

Plant Breeding in the Future

Success in any line of business depends on being able to anticipate and respond to changing market requirements. Such is the long-term nature of developing a new crop variety that plant breeders are in the unenviable position of forecasting their customers' demands at least five, and probably nearer ten years, in advance. Plant breeding objectives will continue to be shaped by changes in agricultural and environmental policy. There is no doubt that while improvements in crop yield and endues quality remain of paramount importance, the future for British agriculture lies in crop production systems which deliver both economic and environmental benefits.

Increasingly, plant breeders are targeting varieties suitable for low input and organic regimes. It is already estimated that genetic improvements in disease resistance alone save British farmers more than £100 million a year in reduced agrochemical use and improved yield. A range of other factors also determines a variety's suitability for low input, integrated farming systems. They include the crop growth cycle, straw strength and susceptibility to weed competition, as well as the variety's ability to thrive in a range of soil types, climatic conditions and rotational patterns.

Novel Crops

Breeders continue to adapt novel crops for UK growing conditions. New varieties of sunflowers and soya beans are already creating market opportunities in the food sector. Post-BSE, attention has focused on using more home-grown fodder in animal feed. For plant breeders, this means enhancing energy and protein levels in traditional forage crops, and developing the potential of new crop species, such as lupins.

Non-food Crops

The use of crop plants for industrial, nonfood applications is an area of increasing interest. Developments in plant breeding could underpin a transition from many crude, industrial processes towards renewable, field-based production of specialist chemicals, textiles and biofuels Quick-growing biomass crops such as willow coppice and miscanthus are already fuelling UK power stations, and there is huge potential to develop oilseed, starch and fibre crops for a wide range of product including lubricants, pharmaceuticals, dyes, plastics, flooring, paper and clothing.

New Traits

Pioneering research by British scientists is opening up new possibilities for crop improvement. Examples from current research include the use of genetic modification to overcome grey mould infestation in strawberries–not possible through conventional breeding because no resistance genes exist in the world strawberry germplasm. Similar techniques offer ways to control the ripening process in fruit and vegetables, to extend storage life, and improve flavour and texture for consumers. It will also increase the seasonal availability of homegrown produce, so reducing the distance food is transported. Modern biotechnology can be used to improve human nutrition. Recent research by British scientists includes the development of vegetables containing increased levels of cancer-fighting nutrients, and apples, which really do keep the dentist at, bay by protecting against tooth decay.

PATENTS

Recent developments in technology have also seen the emergence of patent applications in relation to plant breeding. This is particularly the case where technologies can be applied to several different varieties of the same crop, or across a range of different species. Such developments cannot be protected under the variety-specific system of Plant Breeders' Rights.

The use of patents where living organisms are concerned continues to arouse controversy, but future research will not take place unless some mechanism exists to reward and promote sharing of knowledge.

ISSUES AND CONCERNS

Modern plant breeding, whether classical or through genetic engineering, comes with issues of concern, particularly with regard to food crops. The question of whether breeding can have a negative effect on nutritional value is central in this respect. Although relatively little direct research in this area has been done, there are scientific indications that, by favouring certain aspects of a plant's development, other aspects may be retarded. Reductions in calcium, phosphorus, iron and ascorbic acid were also found. The study, conducted at the Biochemical Institute, University of Texas at Austin, concluded in summary: *"We suggest that any real declines are generally most easily explained by changes in cultivated varieties between 1950 and 1999, in which there may be trade-offs between yield and nutrient content*

The debate surrounding genetic modification of plants is huge, encompassing the ecological impact of genetically modified plants and the safety of genetically modified food.

Plant breeders' rights is also a major and controversial issue. Today, production of new varieties is dominated by commercial plant breeders, who seek to protect their work and collect royalties through national and international agreements based in intellectual property rights. The range of related issues is complex. In the simplest terms, critics of the increasingly restrictive regulations argue that, through a combination of technical and economic pressures, commercial breeders are reducing biodiversity and significantly constraining individuals (such as farmers) from developing and trading seed on a regional level. Efforts to strengthen breeders' rights, for example, by lengthening periods of variety protection, are ongoing.

SPEEDING-UP PLANT-BREEDING WORK

Plant-breeding work involving crosses generally takes a great number of generations before the superior new plant has been obtained. In the routine work of a large plant-breeding station so many breeding projects arc under way simultaneously that it matters little how much time elapses between the initial cross and the final purification of a promising commercial variety. Every year there are many projects being started, several are under way, and a few arc in their last stages.

It often pays to cut down the time required for work of this kind. One of the most promising means of doing this is to grow more than one generation of plants in one year. This usually means that one of those generations of plants is grown in a season in which the plant is not ordinarily grown, and that for this reason a correct evaluation of the quality of the individual plants is out of the question.

In Europe or in America growing an extra generation on the spot means growing plants in the greenhouse. To grow cereals or beans in winter in greenhouse conditions not only requires heat; we must also have the means of regulating the length of day. The simplest method is to use large electric lights to furnish both the heat and the light, to lengthen the day by those means, and to let the temperature go down during the dark part of the night. Under those conditions cereals and beans will flower and set seed very well in the winter months,

The electricity required per square yard is roughly that used by lamps that require a kilowatt. It is clear that under such conditions the winter generation will not show the normal qualities of the plants in that group which are grown in summer-time in the field, but there are many cases in which this is not of much importance.

If we want to proceed by the method of making some hybrids and then growing a second and third inbred generation from this stock, we can do our crossing in the field, grow our hybrids in the winter, and then grow another crop of second-generation plants in normal field conditions. When we are using the method of mating back hybrid stock to some pure kind of plants we want to improve it matters but little if we cannot very well select the best plants during some of the generations involved. It is evident that we must grow a generation under normal field conditions whenever we want to make our selections.

In recent years a few seed-firms have hit upon the perfect scheme of growing two generations of annual plants. This does not involve any greenhouse generations. They work in two different localities with very similar climate, but so chosen that one of those localities is situated in the Southern Hemisphere. There are regions of Chili where Californian firms can find conditions of plant-growth that are identical with conditions at home, with the exception that the growing season comes during the Californian winter. In those circumstances it is quite possible to grow two absolutely normal generations during one year. Our present greatly improved methods of communication make such schemes perfectly feasible.

With many tropical plants one generation yearly is grown of plants that need special conditions of either a wet or a dry season. Here it is easy enough to interpolate an additional generation, by looking for a suitable spot, or by providing the necessary irrigation. Sometimes it takes a very long time to bring the plants to a condition where we can evaluate them. A good example is that of the fruit trees. It takes many years for an apple seedling or a seed-grown cherry to come into bearing. Here we can save a great deal of time by budding or top-grafting parts of the seedlings upon mature trees in full bearing. By this method we may save more than half the number of years normally required for bringing the seedlings to maturity.

In this connection we might also treat of those cases where certain seeds take a very long time to germinate. Storing the seed in suitable conditions of moisture, and especially of temperature, will speed-up germination in such plants as gooseberries and celery. In stone-fruits it has been found possible to crack open the stone, to extract the kernel and grow the seedlings under aseptic or even under antiseptic conditions (thymol solution).

SHOWS AND SHOWING

The exhibiting of plants and seed at the shows is a very curious thing. It pays extremely well for a seed-firm to employ some gardeners who know just exactly when and' how to get enough plants full of luscious beans or covered with flowers on the exact day for each show. Florists and lovers of gardening see the quality of such plants at the shows, and many people who admire them there try in their turn to win the coveted prizes at their local shows the next year. This show game is certainly very nice and pleasant; it gives the gardeners something to strive after; it may even help to keep some of the lads in the country.

On the other hand, there is very, very little connection between the qualities judged at the shows and actual value of the material shown. The few selected mangolds, or the mammoth pumpkins, or the bundle of wheat-ears, or even of the maize-ear contests of the American agricultural shows.

All along the line, in horticulture as well as in agriculture, the selected samples shown give no indication of the value of the seed or plant stock the producers have for sale. Of course we can see that those few mangolds have the ideal shape and size, but the seed-firm must give us some guarantee about the percentage of mangolds that will grow to this size and shape from the seed they sell, just as the owner of a prize-winning bull must give some guarantee of this animal's ability to give profitable daughters, in addition to the blue ribbon awarded to the beautifully coloured and shaped prize-winner.

In agriculture this is beginning to be so well realised that the classes for beautiful corn cobs or for the largest mangolds are becoming very rare indeed, even if we still see stock-judging in dairy cows and similar farces. In the horticultural shows much interest is still shown in regard to flowers and vegetables.

It seems evident that when we strive for beauty, beauty contests for roses and begonias, for asparagus and delphiniums and dahlias are indicated. The show is a shop-window. The seller must show if his competitors show, if only to keep his name before the public. To the public the show is only just a pleasant occasion for a half-day outing. To the buyer the show is much nicer to see than a description in a catalogue.

Long lists of novelties are bought at the shows by people with an experimental turn of mind. We see the representatives at the shows kept busy noting down long lists of orders. Showing certainly pays the seller. A very much more sensible system of showing a great many varieties is that of the demonstration gardens and experimental stations, both stations in which the agricultural authorities try to compare promising new and good old varieties with the aid of objective standards, and such gardens as show-plots arranged by rose-lovers or by societies of growers of gladioli or dahlias. Here the varieties can be judged throughout the growing season in favourable conditions.

THE ANNUAL SELF-FERTILIZING CROPS

The number of plant species now being cultivated by man is really enormous, while only relatively few animal species are kept under domestication. It is possible for the author of a book on animal breeding to take those species almost one by one, and to give the specialist breeders a few hints over and above those which are contained in the general book. This would be quite out of the question in a book on plant breeding. One way out of the difficulty would be to give one example of each of the three groups—self-fertilized plants, cross-fertilizers and plants that are usually propagated by vegetative means. These plants in two sections— grains and legumes—giving some hints to the breeders of wheat, barley, oats, sorghum, rice, etc., and going into some details about the breeding of beans, soybeans, cow-peas, ground-nuts, peas, etc., separately. But, after all, the self-fertilizing annuals have so much in common from the point of view of plant-breeding methods.

Self-fertilization causes isolation. Each plant, as far as reproduction is concerned, is as self-contained and isolated in a crowded field as it would be if grown by itself in a greenhouse in a city. The inbreeding caused by self-fertilization has several results. It makes all heritable variability disappear in every line, it makes every group consist of parallel lines, in which each plant descends from only one parent plant, so that the usual complicated network of ancestry is simplified into a system of isolated parallel lines. Where inherited variability exists, such as happens after a deliberate artificial cross or when accidental cross-fertilization occurs, it will automatically disappear.

That is to say, it will disappear in so far as each separate line is concerned, for Mendelian segregation will make the lines, descending from any hybrid heterozygous for a great many genes, diverse in their hereditary make-up. The result is that a field of self-fertilizing annuals that has not been grown from just one pure line by a plant breeder always consists of a mixture of a number of genetically different lines. Even if accidental cross-fertilization or irregularities in crossing-over (mutation) are very rare, they do occur, and cause this state of things. This genetic variability, however, is wholly different from what we find in other plants and animals. We are dealing with mixtures of pure lines, and almost every plant we take from any field is pure, homozygous for all its genes, and will, if its seeds are sown, give a very homogeneous descendance. From all this it follows that enormous progress can often be made in all self-fertilized plants by simply picking out a great number of promising-looking individual plants and comparing their descendance.

Many valuable widely grown wheats, tobaccos, peanuts and beans have been just found. They "just growed ", like Topsy. The first step in plant-breeding in the self-fertilizing plants must always be to look for those ready-made pure lines. The next is to find whether a mixture of two of the very best pure lines can be discovered which will give us still higher production (10 per cent, higher production from a mixture of two strains as compared with that of the best one is quite common).

We have then to produce some more genetic variability by deliberate cross-breeding. This generally means that we try to combine the good qualities of two different strains in one new one. Here there are two wholly different schools of thought, and two wholly different methods. One of them is to analyse second-generation plants, and to find those individuals that do combine, let us say, the disease resistance of line A with the large quantity of big seeds of line B. This is a method in- vented by the experimental geneticists who are trying to combine practical plant breeding with purely scientific gene-analysis. As a theoretical geneticist, interested in genes and their action.

It is great fun playing with a hundred different hybrid lots in one species, growing a few dozen second-generation plants in each lot and tabulating the results. But this is neither sound genetics nor sound plant breeding. We are dealing with so many unanalysed and unanalysable genes which all affect production and quality at the same time, that the chance of finding something really worth while in a few hundred plants is simply not good enough.

The other method is based upon the realization that hundreds of genes are commonly involved, and upon the fact that purification—reduction of the potential variability—is rapid and automatic in this material. It consists of growing the greatest possible number of seeds from first-generation hybrid plants. We can then proceed in two different ways. One is an analytical method, which means that we are seeking for the very best-looking plants (for instance, by weighing their seeds), and progeny testing those plants by sowing a large number of seeds from each, in order to find the most profitable families. The other method is what Baur and Nilsson-Ehle called a "Ramsch" method.

This consists simply of growing the descendance of a hybrid in a vast mixture, refraining from any analysis, and continuing this hybrid lot as a miscellaneous collection for five or six generations, sowing as much of the mixed seed of each harvest as we have room for. If we do this, natural selection will weed out most of the unproductive and undesirable types, such as those plants t1 at ripen after harvesting time, all weaklings and all plants susceptible to diseases and pests. After five or six years that lot will consist of a mixture of pure lines, every line different from every other, with the successful lines represented by more individuals than the unsuccessful ones. After this we will treat the mixture just as if we were dealing with any other mixture, growing separate rows of beds each from one likely-looking plant.

This is a very sound method, which saves a great deal of un- necessary labour and time. It is extremely difficult to find just how much better this method is as compared with that of analysis in every generation, starting from the second generation. An analytical method could be made to work very well, provided we did all we could to make the number of plants of the second and third generations exceedingly large (in the cereals we can get almost as many seeds as in tobacco by splitting up the hybrid plant repeatedly before it starts to grow any stalks).

In actual practice there are only a few crosses that really produce anything worth while, and the general practice seems to be to make a great many attempts and grow a large number of first-generation hybrids. This renders it imperative to cut down on the number of plants per experiment. It is a vicious circle. This is all very wrong and wasteful. One cross every year and work with it, both analytically and by means of a "Ramsch ", doing the final selection after six years for one lot in every season, than waste my time castrating flowers and producing numerous hybrids that would have no really good chance of showing what they could do.

In all the self-fertilizing annuals we must be continually on the look-out for qualities that may be of special merit. Mostly of adaptations to special conditions. In wheaa on irrigated ground, resistance to drowning may be as important as disease resistance or quality of the grain. Conversely, rice could be grown much more extensively if we would take the trouble to do some plant-breeding work on dry-land rices, which abound in the mountain villages of all the tropical Asian islands. In ground-nuts we should look for lines that are specially adapted to mechanical harvesting, as well as for lines adapted to the production of a first-quality product in the hands of tens of thousands of small cultivators. In all the plants of this group it is possible not only to improve the yield and the quality, but also to extend the area in which profitable production is possible. Tobacco, rice, soybeans and cow-peas are just as important in this respect as wheat, barley and oats.

Plant breeding with self-fertilizing plants is comparatively easy. It is even possible for the real amateur plant breeder to do some extremely interesting and useful work with some of the less well-known plants of this group—let us say with beans or cow-peas. Such work needs room, but very little special technical knowledge or talent for higher arithmetic. Frequently a few plants will stand out from an otherwise quite even field of barley, beans or sorghum. Such "high birds "—climbing plants in bush beans, high, scmiawned plants in wheat—are often accidental hybrids. If we sow a few of their seeds, we can see immediately whether they are hybrids or admixtures, for hybrids will have very variable offspring. If then we sow all the seeds from such hybrids, we can start a series of experiments that may easily give us something worth while, even when the father of the hybrid is unknown.

7

Evaluation of Plant Breeding

PLANT BREEDER

The aim of every plant breeder is to produce a group that will be worth a place in the assortment of cultivated plants. In some cases the new plant will have only ornamental value, in others it will, for any of a dozen reasons, be better adapted to agriculture than other similar ones. It is easy enough in all classes of plants to produce novelties, to produce variability. The experimental grounds of the plant breeder would soon be cluttered up with novelties if he did not keep on weeding out the comparatively worthless. After producing our novel plants, we must judge them. Some qualities are much easier to judge than others. It is comparatively easy to weed out a lot of seedling dahlias and to pick out a few that may be worth keeping on for further testing. In the selection of our agricultural plants, in cereals, in beets, in maize, the assortment of valuable kinds is already so great that only relatively few of the newer selections are worth keeping. In the vegetatively propagated plants this work of testing novelties is comparatively easy.

In the first place, we are not bothered by considerations of purity—once we have produced a very good specimen, we can just multiply it with all its properties intact. In the second place, we can easily, for purposes of testing and comparison, produce as many individual plants of the novelty as we require. When the differences are relatively small—as in cereals and beets, where the novel plant produced must compete with excellent old stock that has been made by a process of selection and comparison during centuries—a correct system of evaluation becomes of very great importance. What we really want to test is the inherited composition of the material.

This may be difficult, as the final quality of the plants to a large extent depends on favourable or unfavourable conditions under which they grew up. It will perhaps be as well to begin with the simplest problems. In the production of novelties in such flowers as roses or dahlias the value of a new variety depends entirely on the appeal it makes to the gardening public. The man who is proud to have found a novel dahlia seedling or a new and strikingly beautiful rose sends it to a show and hopes that it will take the judge's eye and win him the much- "prized ribbon or certificate. But of course the insider knows that many highly commended novelties will never be heard of again, or that they will drop out of cultivation in a few years. There is a difference between the fact that the grower can bring half a dozen perfect blooms to the show on time, and the question whether that same rose or chrysanthemum will ever be popular in the gardens, or even be worth while growing for the commercial producer of cut flowers. A dahlia that wins the medal may be a very weak grower, or it may be very susceptible to rain.

The strikingly beautiful rose may be very susceptible to mildew, or its flowering season may be extremely short. If we visit the growers of greenhouse roses, or the gardens of the man who produces cut flowers for the market, we find that the number of varieties that have real commercial value—varieties "with bread in them ", as our gardeners express it—is extremely small. We shall probably find only three or four of the relatively older dahlias worth growing for cut flowers, and it is seldom that a novelty is added to that list.

In testing out agricultural plants, both during the process of selection and after they have finally been introduced, many methods are in use. The direct, simple examination of a number of small plots or rows will show us the qualities of those lots, but this method will only serve to find the groups with definitely inferior quality. It is not good enough to find exceptional merit, as circumstances may have favoured those qualities we appreciate. The only way to avoid reaching the wrong conclusions is to see to it that environmental conditions do not vitiate our results. The best way to do this is to com- pare the quality of a number of plants with that of other plants that were grown in wholly comparable conditions.

One of the best ways to do this is to grow the plots of our plant between similar plots of an older variety, that can be used as a standard of comparison. The choice of this standard is of great importance, and we will revert to this subject when treating of the breeding of special plant groups. As an example we can here take the evaluation of a newly produced (or of an imported) potato clone, or pure line of wheat or barley, In a case like this the standard may be some line or clone that is of commercial importance, against which the tested plant will be called upon to compete. If there is sufficient material, it is wise to repeat the comparison—that is to say, we should grow the test plot between standard plots in different parts of the field, or preferably in different fields. To equalize accidental conditions in small parts of the field, oblong plots are generally better than square ones. If we are testing a number of experimental lots, we can grow them side by side, and interpolate standard plots to be used for purposes of comparison.

The experimental garden will need an efficient system of book-keeping. We must note down anything of value about the experimental and the standard plots, date of maturity, disease resistance, quality, yield. We can calculate average yield for the sum of the plots of one kind, and then compare those figures with the average computed yield of the standard plots or with the average yield of the whole field; and a comparison of the deviation of the yields in the different plots with calculated mean errors will give us valuable data about the question in how far the superiority or inferiority of each separate number is significant as an indication of merit.

The practice of plant breeding in different parts of the world. A multitude of miserably small checkerboard fields being weighed at harvest time to produce the figures for an office staff to juggle with. A successful establishments where the field-work was under the supervision of men who were thoroughly conversant with the material, and where excellent results were obtained with a minimum of calculating machine work. A method of annotating the results obtained in experimental plots which gives the plant breeder an excellent insight into the merits of his new material is that of graphical registration. This method can be adapted to the most diverse projects; it is very simple, and gives us our results at a glance. The method consists of registering the quality (yield, height, resistance) of each plot, both standard plots and testing plots, on ruled paper.

If we do this, the line that can be drawn through the points of the standard plots will probably rise in one part of the field, and descend in some other part of it, and by drawing a line that joins those points we obtain a basis upon which we can found our judgement of the results obtained in the experimental plots. If we want to test our experimental material in two or three different qualities, we must construct separate graphs for each of those qualities, or sometimes we can use the same graph, employing different colours for separate qualities (weight, sugar percentage, purity in beets, etc.).

Reading those graphs will now show us the quality of our plant plots, in comparison with the standards, and we can use this relative value as a basis for our selection work. If we compare half a dozen groups in one experiment, and we find that one or two of them will in every duplication remain well below the standard in quality, this may be a very good reason for discarding them. On the other hand, if we find one or two numbers that all through the experiment are well above the standard line in quality, we have a very good reason for preferring them. It may be that in the case of one or two numbers the results are very variable, so that in one part of the field the number is above and in other parts it is below the standard in quality. Then we may well retain this number for another trial in the next season.

The influence of one row of plants on those growing close to it may be considerable, and we must guard against this. If we can grow three rows and use only the middle one for our records, we take a step in the right direction. If we have enough seed to plant six rows, and we examine only the two middle ones, discarding two rows on each side, this is still better. The greatest difficulty arises in those cases where we are dealing with very few plants of each number; or, worse, where, as in the second generation of a cross, every plant is probably different from every other. It then becomes very important to equalize conditions for each plant as far as this is possible. A very good plan is to grow the plants in rows, and alternate those experimental rows with rows of some pure standard variety. Each plant may then have two diverse neighbours in its own row, but it will touch four other plants that will be of the same kind as four out of six neighbour plants in the case of every other plant in the experiment.

This means that to save time we must make a preliminary choice out of the many numbers in those early generations, and it is very important that we shall make the cultural conditions as nearly equal as possible for every plant. This is the place to warn against using apparent correlative qualities in judging the value of our experimental plants. In the breeding of animals even to-day we often see that almost as much attention is paid to correlative qualities as to real merit.

Some of this is certainly due to the fact that in some material real merit can only be judged in one of the two sexes. There is no good reason to give any attention to correlative qualities. In so far as a real correlation exists, selection for real quality will automatically bring along the correlative character, and in as far as we are mistaken (and this will nearly always be so) we are simply led astray by paying attention to apparent correlations. In plant-breeding selection according to qualities believed to be indirect indications of good quality used to be very frequently attempted.

This has gone out of fashion, and we only meet traces of it in very special fields (sugar-cane, apples). In our selection work we should always consider actual practical merit, and we should care- fully study in how far the qualities we can examine, measure and weigh are indications of this real merit.

CHOICE OF PLANT-BREEDING MATERIAL

The advantage of having at one's disposal a large collection of plant varieties, both plants of commercial importance and imported plants of no immediate practical importance, is enormous. A difficult question for the plant breeder is always what crosses to make, what plants to combine with the greatest hope of ultimate success.

In the breeding of such plants as potatoes, raspberries and roses, certain clones have the reputation of being profitable parents of hybrid lots. The Early Rose potato, the Lloyd George raspberry, the Chilean strawberry. In some special instances the choice is obvious. When our present assortment of flaxes or potatoes or roses happens to be deficient in some such quality as disease resistance or winter- hardiness, we are on the look-out for plants that will show those desirable qualities to the greatest extent. In other words, we shall, when choosing the parents of our future hybrids, try to find material that promises the possibility of a combination of the very good points of both.

This is obvious. Often crosses that seemed very little promising have given excellent results. It would seem that here no definite rule could be indicated. It is always worth while trying crosses with new material, with newly imported species—in fact with material which has not been used previously. This is especially so when, as in horti- culture, everybody is on the look-out for something new. Another rule that could be given would be that when we make very wide crosses we cannot hope to find commercially valuable strains in the second hybrid generation, unless we grow extremely large numbers of plants in that generation. When we make a very wide cross to incorporate some special quality into the existent stocks, the best thing we can do is to use the hybrids in back-cross series.

We can then strive to retain just one special quality of the new species and incorporate it into the general set of genes with which we are familiar in our older stock. In several instances very wide crosses have been successfully used, and sometimes plants have been used in cross-breeding that in themselves showed very little promise. The inedible cattle gourd that Orton crossed with his water- melons, of some of the potato crosses with Andean wild species (Salaman) and of the back-crosses to hybrids with wild cane from which all our modern resistant varieties of sugar-cane have been derived.

It is here—in the choosing of plant material for breeding work—that cytological studies and chromosomal counts are of the greatest practical value.

BREEDING METHODS IN CROP PLANTS

SELF POLLINATED CROPS

Mass selection: In mass selection, seeds are collected from (usually a few dozen to a few hundred) desirable appearing individuals in a population, and the next generation is sown from the stock of mixed seed. This procedure, sometimes referred to as phenotypic selection, is based on how each individual looks. Mass selection has been used widely to improve old "land" varieties, varieties that have been passed down from one generation of farmers to the next over long periods.

An alternative approach that has no doubt been practiced for thousands of years is simply to eliminate undesirable types by destroying them in the field. The results are similar whether superior plants are saved or inferior plants are eliminated: seeds of the better plants become the planting stock for the next season.

A modern refinement of mass selection is to harvest the best plants separately and to grow and compare their progenies. The poorer progenies are destroyed and the seeds of the remainder are harvested. It should be noted that selection is now based not solely on the appearance of the parent plants but also on the appearance and performance of their progeny. Progeny selection is usually more effective than phenotypic selection when dealing with quantitative characters of low heritability. It should be noted, however, that progeny testing requires an extra generation; hence gain per cycle of selection must be double that of simple phenotypic selection to achieve the same rate of gain per unit time.

Mass selection, with or without progeny test, is perhaps the simplest and least expensive of plant-breeding procedures. It finds wide use in the breeding of certain forage species, which are not important enough economically to justify more detailed attention.

Pure-line selection: Pure-line selection generally involves three more or less distinct steps: (1) numerous superior appearing plants are selected from a genetically variable population; (2) progenies of the individual plant selections are grown and evaluated by simple observation, frequently over a period of several years; and (3) when selection can no longer be made on the basis of observation alone, extensive trials are undertaken, involving careful measurements to determine whether the remaining selections are superior in yielding ability and other aspects of performance.

Any progeny superior to an existing variety is then released as a new "pure-line" variety. Much of the success of this method during the early 1900s depended on the existence of genetically variable land varieties that were waiting to be exploited. They provided a rich source of superior pure-line varieties, some of which are still represented among commercial varieties. In recent years the pure-line method as outlined above has decreased in importance in the breeding of major cultivated species; however, the method is still widely used with the less important species that have not yet been heavily selected.

A variation of the pure-line selection method that dates back centuries is the selection of single-chance variants, mutations or "sports" in the original variety. A very large number of varieties that differ from the original strain in characteristics such as colour, lack of thorns or barbs, dwarfness, and disease resistance have originated in this fashion.

Hybridization

During the 20th century planned hybridization between carefully selected parents has become dominant in the breeding of self-pollinated species. The object of hybridization is to combine desirable genes found in two or more different varieties and to produce pure-breeding progeny superior in many respects to the parental types.

Genes, however, are always in the company of other genes in a collection called a genotype. The plant breeder's problem is largely one of efficiently managing the enormous numbers of genotypes that occur in the generations following hybridization. As an example of the power of hybridization in creating variability, a cross between hypothetical wheat varieties differing by only 21 genes is capable of producing more than 10,000,000,000 different genotypes in the second generation. At spacing normally used by farmers, more than 50,000,000 acres would be required to grow a population large enough to permit every genotype to occur in its expected frequency. While the great majority of these second generation genotypes are hybrid (heterozygous) for one or more traits, it is statistically possible that 2,097,152 different pure-breeding (homozygous) genotypes can occur, each potentially a new pure-line variety. These numbers illustrate the importance of efficient techniques in managing hybrid populations, for which purpose the pedigree procedure is most widely used.

Pedigree breeding starts with the crossing of two genotypes, each of which have one or more desirable characters lacked by the other. If the two original parents do not provide all of the desired characters, a third parent can be included by crossing it to one of the hybrid progeny of the first generation. In the pedigree method superior types are selected in successive generations, and a record is maintained of parent–progeny relationships.

The F2 generation (progeny of the crossing of two F1 individuals) affords the first opportunity for selection in pedigree programmes. In this generation the emphasis is on the elimination of individuals carrying undesirable major genes. In the succeeding generations the hybrid condition gives way to pure breeding as a result of natural self-pollination, and families derived from different F2 plants begin to display their unique character. Usually one or two superior plants are selected within each superior family in these generations. By the F5 generation the pure-breeding condition (homozygosity) is extensive, and emphasis shifts almost entirely to selection between families.

The pedigree record is useful in making these eliminations. At this stage each selected family is usually harvested in mass to obtain the larger amounts of seed needed to evaluate families for quantitative characters. This evaluation is usually carried out in plots grown under conditions that simulate commercial planting practice as closely as possible. When the number of families has been reduced to manageable proportions by visual selection, usually by the F7 or F8 generation, precise evaluation for performance and quality begins. The final evaluation of promising strains involves (1) observation, usually in a number of years and locations, to detect weaknesses that may not have appeared previously; (2) precise yield testing; and (3) quality testing. Many plant breeders test for five years at five representative locations before releasing a new variety for commercial production.

The bulk-population method of breeding differs from the pedigree method primarily in the handling of generations following hybridization. The F2 generation is sown at normal commercial planting rates in a large plot. At maturity the crop is harvested in mass, and the seeds are used to establish the next generation in a similar plot. No record of ancestry is kept.

During the period of bulk propagation natural selection tends to eliminate plants having poor survival value. Two types of artificial selection also are often applied: (1) destruction of plants that carry undesirable major genes and (2) mass techniques such as harvesting when only part of the seeds are mature to select for early maturing plants or the use of screens to select for increased seed size. Single plant selections are then made and evaluated in the same way as in the pedigree method of breeding.

The chief advantage of the bulk population method is that it allows the breeder to handle very large numbers of individuals inexpensively.

Often an outstanding variety can be improved by transferring to it some specific desirable character that it lacks. This can be accomplished by first crossing a plant of the superior variety to a plant of the donor variety, which carries the trait in question, and then mating the progeny back to a plant having the genotype of the superior parent. This process is called backcrossing. After five or six backcrosses the progeny will be hybrid for the character being transferred but like the superior parent for all other genes. Selfing the last backcross generation, coupled with selection, will give some progeny pure breeding for the genes being transferred. The advantages of the backcross method are its rapidity, the small number of plants required, and the predictability of the outcome. A serious disadvantage is that the procedure diminishes the occurrence of chance combinations of genes, which sometimes leads to striking improvements in performance.

Hybrid Varieties

The development of hybrid varieties differs from hybridization. The F1 hybrid of crosses between different genotypes is often much more vigorous than its parents. This hybrid vigour, or heterosis, can be manifested in many ways, including increased rate of growth, greater uniformity, earlier flowering, and increased yield, the last being of greatest importance in agriculture.

CROSS POLLINATED CROPS

The most important methods of breeding cross-pollinated species are (1) mass selection; (2) development of hybrid varieties; and (3) development of synthetic varieties. Since cross-pollinated species are naturally hybrid (heterozygous) for many traits and lose vigour as they become purebred (homozygous), a goal of each of these breeding methods is to preserve or restore heterozygosity.

Mass Selection

Mass selection in cross-pollinated species takes the same form as in self-pollinated species; *i.e.,* a large number of superior appearing plants are selected and harvested in bulk and the seed used to produce the next generation. Mass selection has proved to be very effective in improving qualitative characters, and, applied over many generations, it is also capable of improving quantitative characters, including yield, despite the low heritability of such characters. Mass selection has long been a major method of breeding cross-pollinated species, especially in the economically less important species.

Hybrid Varieties

The outstanding example of the exploitation of hybrid vigour through the use of F1 hybrid varieties has been with corn (maize). The production of a hybrid corn variety involves three steps: (1) the selection of superior plants; (2) selfing for several generations to produce a series of inbred lines, which although different from each other are each pure-breeding and highly uniform; and (3) crossing selected inbred lines. During the inbreeding process the vigour of the lines decreases drastically, usually to less than half that of field-pollinated varieties. Vigour is restored, however, when any two unrelated inbred lines are crossed, and in some cases the F1 hybrids between inbred lines are much superior to open-pollinated varieties. An important consequence of the homozygosity of the inbred lines is that the hybrid between any two inbreds will always be the same. Once the inbreds that give the best hybrids have been identified, any desired amount of hybrid seed can be produced.

Pollination in corn (maize) is by wind, which blows pollen from the tassels to the styles (silks) that protrude from the tops of the ears. Thus controlled cross-pollination on a field scale can be accomplished economically by interplanting two or three rows of the seed parent inbred with one row of the pollinator inbred and detasselling the former before it sheds pollen. In practice most hybrid corn is produced from "double crosses," in which four inbred lines are first crossed in pairs (A × B and C × D) and then the two F1 hybrids are crossed again (A × B) × (C × D). The double-cross procedure has the advantage that the commercial F1 seed is produced on the highly productive single cross A × B rather than on a poor-yielding inbred, thus reducing seed costs. In recent years cytoplasmic male sterility, described earlier, has been used to eliminate detasselling of the seed parent, thus providing further economies in producing hybrid seed.

Much of the hybrid vigour exhibited by F1 hybrid varieties is lost in the next generation. Consequently, seed from hybrid varieties is not used for planting stock but the farmer purchases new seed each year from seed companies.

Perhaps no other development in the biological sciences has had greater impact on increasing the quantity of food supplies available to the world's population than has the development of hybrid corn (maize). Hybrid varieties in other crops, made possible through the use of male sterility, have also been dramatically successful and it seems likely that use of hybrid varieties will continue to expand in the future.

Synthetic Varieties

A synthetic variety is developed by intercrossing a number of genotypes of known superior combining ability—*i.e.,* genotypes that are known to give superior hybrid performance when crossed in all combinations. (By contrast, a variety developed by mass selection is made up of genotypes bulked together without having undergone preliminary testing to determine their performance in hybrid combination.)

Synthetic varieties are known for their hybrid vigour and for their ability to produce usable seed for succeeding seasons. Because of these advantages, synthetic varieties have become increasingly favoured in the growing of many species, such as the forage crops, in which expense prohibits the development or use of hybrid varieties.

BREEDING PERENNIAL GRAINS: COOL-SEASON GRASSES

WHEAT

Development of perennials up to 1990: Hexaploid wheat arose approximately 5000 years ago when the genomes of tetraploid wheat (*T. turgidum* or *T. carthlicum,* both 2n=28, AABB) and an Asian goatgrass (*Aegilops tauschii,* 2n=14, DD) were combined via amphiploidization (*i.e.,* natural hybridization followed by spontaneous production of 2n gametes in the hybid.) Tetraploid wheat itself is a much older natural amphiploid, incorporating the genomes of two diploid grasses. Both wheats are part of the large and diverse Triticeae tribe of the grass family (Gramineae), which also includes scores of perennial species, many of which can be hybridized with wheat. Wagoner (1990a) examined in detail the early history of efforts in the United States, Canada, Germany, and, most importantly, the USSR, to transfer genes for perenniality from alien grass species into bread wheat, citing more than 65 publications on the subject. None of these efforts produced a truly perennial grain cultivar, but they did spin off much valuable annual germplasm with genes for disease resistances and other traits. In the end, most of the effort in the perennial-wheat programmes was diverted into producing improved annual cultivars, where progress was more easily achieved.

Of the few perennial, grain-producing genotypes developed from wide hybrids at the time of Wagoner's review, none was agronomically successful. Soviet-developed 'perennial' cultivars produced good grain harvests only in the year in which they were established from seed; in the end, they were used mainly as forage cultivars that provided no more than one grain harvest. The US germplasm 'MT-2', derived from a hybrid between *T. turgidum* and

Thinopyrum intermedium (2n=42), released by Schulz-Schaeffer and Haller (1987) in Montana, had very low kernel weight and unreliable persistence. In Sweden, Fatih (1983) found that yields of perennial *T. aestivum*/*Th. intermedium* partial amphiploids (2n=56) were, on average, only 48% of the yields of 42-chromosome, annual, backcross-derived lines of similar parentage.

Production of new hybrids: In the decade since Wagoner's review, no perennial wheat cultivars have been released for production. But basic research on hybridization and cytogenetics has opened up new possibilities for geneticists and breeders interested in the problem.

Species of the genus *Thinopyrum* have been hybridized with wheat more often than have any other perennial species because of the ease of producing partially fertile hybrids, often without embryo rescue. *Thinopyrum intermedium* (2n=42) is rhizomatous, and two other commonly utilized species, *Th. ponticum* (2n=70) and *Th. elongatum* (2n=14) are caespitose. Researchers at Washington State University have launched a new programme to develop perennial wheat from *Thinopyrum* crosses, with promising preliminary results. As we shall, recent cytological and molecular studies have explained why *Thinopyrum* crosses have often led to frustration, and in doing so, have suggested new approaches.

Other species in the genera *Thinopyrum, Elymus,* and *Leymus* have long been investigated as sources of perenniality, and the perennial gene pool available to wheat geneticists is growing rapidly. In a review, Sharma and Gill (1983) listed only 16 perennial species that had been hybridized with hexaploid or tetraploid wheat, including only one species — *Elymus giganteus* — outside the *Agropyron-Thinopyrum* complex. A decade later, Jiang *et al.* (1994) added 38 additional species, including 17 in *Elymus* and 6 in *Leymus,* to the list of hybrids. Sharma (1995) reviewed the production of hybrids between wheat and more than 50 perennial species.

Our ability to make crosses and backcrosses between distantly related species has grown along with advances in embryo rescue, hormone treatments, intra-ovarian fertilization, bridge crosses, and protoplast fusion. The effects of these techniques are magnified when they are used to exploit intraspecific variation for crossability, combining ability, or variation among species carrying the same or similar genomes. Especially important are reciprocal crosses followed by embryo rescue when the traditional method of using the species with higher chromosome number fails.

Within each perennial species, accessions can vary in crossability, and choice of the annual wheat parent can also have a strong effect. For example, homozygosity for kr crossability alleles often makes hybridization possible, and may even improve early seed development; however, it may not affect results in extremely wide crosses. Choice of the wheat parent may depend on other considerations. Hybrids of *Thinopyrum* and *Leymus* with Chinese Spring (*krkr*) could not survive winter temperatures, whereas hybrids with two

Japanese spring wheats were winterhardy. Fertility in the F1 through production of unreduced gametes can be induced by crossing the perennial species with the tetraploid wheat *T. carthlicum* and maintaining low temperatures during pollination and embryo development.

Many wheat/*Elymus* hybrids have been made in recent years, and addition, substitution, and translocation lines have been developed. But to our knowledge, no explicit effort to transfer perenniality from *Elymus* to wheat is underway. Lu and von Bothmer (1991) produced hybrids between 12 *Elymus* species (2n=28 or 42) and wheat (2n=42). Hybridization was made possible by using *Elymus* as the female parent and rescuing hybrid embryos, as first demonstrated by Sharma and Gill (1983). All hybrids were perennial, but chromosome pairing was very low, as expected. The hybrids were not treated with colchicine to produce amphiploids.

Hybridization between wheat and the genus *Leymus* has a long history, but has not led to perennial cultivars. Hybrids are most easily obtained with the larger-seeded, self-pollinated species of the genus: *L. arenarius* (2n=56), *L. racemosus* (2n=28), and *L. mollis* (2n=28) (Dewey, 1984), and these species also have the greatest agricultural potential. Wheat has been crossed with eight *Leymus* species in all. Most crosses require embryo rescue. Hormonal treatment has been used in producing F1 seed.

Wheat/*Leymus* hybrids produced by the Soviet perennial-wheat breeding programme were perennial but less winterhardy than wheat-*Thinopyrum* hybrids. In recent studies, hybrids produced by pollinating two wheat species (*T. aestivum* and the tetraploid *T. carthlicum*) with *L. arenarius* and *L. mollis* were perennial, producing short rhizomes and exhibiting some intergenomic pairing and a high enough level of fertility to permit backcrossing. Hybrids between *T. aestivum* and two other species, *L. innovatus* and *L. multicaulis,* were non-rhizomatous.

Advances in chromosome identification: A revolution in cytogenetic techniques has led to a better understanding of the problems faced by perennial wheat breeders. MT-2 provides a good example. It was derived by selfing a 70-chromosome amphiploid containing the genomes of durum wheat (*T. turgidum,* 2n=28, genomes AABB) and *Th. intermedium,* (2n=42, genomes $StStEEE^{St}E^{St}$). Schulz-Schaeffer and Haller (1987) predicted that MT-2 would stabilize at 2n=56 through elimination of *Thinopyrum* chromosomes. In fact, according to Jones *et al.* (1999), individual plants of MT-2 vary in chromosome number, with most having 2n=56. But genomic *in situ* hybridization (GISH) showed that 56-chromosome MT-2 plants contained numbers of wheat chromosomes varying between 24 and 28. Therefore, there had been loss of wheat chromosomes in some plants, and *Thinopyrum* chromosomes had been eliminated at random. Plants contained up to four St-E or St-E^{st} translocated chromosomes but no wheat-*Thinopyrum* translocations. A wheat/*Thinopyrum* hybrid known as AT 3425, which has resistance to Cephalosporium stripe disease and perennial growth habit, has 2n=56; fluorescent genomic in situ hybridization (FGISH) and C-banding showed that AT 3425 has 36 wheat chromosomes, 14 *Thinopyrum* chromosomes, and 6 chromosomes resulting from translocations between the two species. The *Thinopyrum* chromatin in AT 3425 and another line with the same chromosome configuration, PI 550713, probably originated from *Th. ponticum,* and both lines are cytologically stable. Another perennial, 56-chromosome line, AgCs, carries the combined genomes of hexaploid wheat and the diploid species *Th. elongatum.*

Banks *et al.* (1992) examined meiotic pairing in crosses among eight 56-chromosome partial amphiploids derived from crosses between hexaploid wheat (*T. aestivum*) and *Th. intermedium.* As is often found, the lines had all 42 wheat chromosomes as well as 14 from *Thinopyrum.* The sets of chromosomes originating from *Th. intermedium* differed in all but two lines. Unfortunately, all eight amphiploids studied meiotically by Banks *et al.* (1993) were annual; two Soviet-developed perennial wheats were examined only phenotypically.

Recently, partial amphiploids that had originated in the Soviet progam have been found to be hexaploid, containing 30 chromosomes from tetraploid wheat and 12 from *L. mollis.* Thus, as in wheat-*Thinopyrum* amphiploids, elimination of the perennial parent's chromosomes had occurred. In addition, one pair of wheat chromosomes had been substituted for a pair from *L. mollis.* The partial amphiploids were annual.

Thinopyrum elongatum (2n=14) is a diploid that is more difficult to hybridize with wheat than are *Th. intermedium* or *Th. ponticum;* however, the resulting amphiploids are vigorous, stable, and perennial. Because *Th. elongatum* contributes only a single genome, all plants should have the same 56-chromosome complement. For example, the wheat/*Th. elongatum* amphiploid AgCs is cytologically stable and perennial. Jauhar (1992) produced trigeneric hybrids between a *Th. bessarabicum-Th. elongatum* amphiploid and tetraploid wheat (*T. turgidum*). The hybrids (2n=28, ABJE) were vigorous and perennial.

The maximum number of chromosomes that can be tolerated in amphiploids between wheat and *Thinopyrum* spp. appears to be 56, although 42-chromosome genotypes are more meiotically stable. Usually, partial amphiploids resulting from crosses with tetraploid wheat will contain approximately 28 chromosomes from the wheat parent and approximately 28 from *Thinopyrum,* whereas partial amphiploids with hexaploid wheat will have approximately 42 from wheat and 14 from *Thinopyrum.* Cauderon (1979) pointed out that partial amphiploids are automatically selected for "good balance" between wheat and perennial chromosomes through the elimination process; however, as we have seen, the same chromosomal complements are not consistently selected. Breeding populations based on collections of such lines would be plagued with sterility and lack of chromosome pairing unless a diploid perennial parent such as *Th. elongatum* is used.

The goal of a breeding programme cannot be to develop a single perennial wheat cultivar. One partial amphiploid carefully selected to be cultivated as a perennial would have a unique chromosomal constitution. It would be a gene pool of one individual — a dead end in a breeding programme. To launch a perennial wheat breeding programme based on partial amphiploids would be an ambitious undertaking, involving the following steps for any polyploid perennial species targeted:

- Hybridize tetraploid and hexaploid wheats with the polyploid perennial species, making many parental combinations and sampling diversity of all parental species.
- Produce amphiploids and self-pollinate with mild selection for enough generations to achieve stable chromosome numbers.
- Use *in situ* hybridization, chromosome banding, genetic markers, and other techniques to identify chromosomes in a large population of selected plants representing many parental combinations.
- Assign partial amphiploids to groups of homogeneous chromosomal constitution.
- Compare groups for all phenotypic traits of interest and select one or a few on which to base further breeding. Develop foundation breeding pools from those groups. Experiment with intercrosses between groups, selecting for potentially superior chromosomal combinations.
- In creating new partial amphiploids to introduce into breeding pools, select strictly for appropriate chromosomal complements.

Such a plan would require a mammoth investment of resources, but the initial production of stable partial amphiploids is feasible on a large scale. Selfing and stabilization will occupy several years; by the time genotypes requiring chromosome identification can be produced, vastly more efficient cytological and molecular techniques are almost certain to be available, bringing the breeding programme into the realm of the practical. But technological improvements do not guarantee the development of truly perennial grains, and parallel strategies are needed.

New strategies for perennial wheat: The perennial grasses of the tribe Triticae have long been used in wheat improvement, primarily as sources of individual resistance genes. A traditional strategy for transferring genes is to produce F1 hybrids between wheat and the donor species; either double the chromosome number of the hybrid to produce an amphiploid or pollinate the F1 directly; backcross to wheat genetic stocks to produce lines carrying the normal wheat complement of 42 chromosomes plus one or a pair of chromosomes from the donor parent; and — at some point in the process — attempt to induce a translocation that transfers a segment carrying the target gene to a wheat chromosome. There are many variations on this strategy, but the usual goal is to transfer a single gene, eliminating as much of the rest of the donor genome as possible. Perenniality in wheat's relatives is more genetically complex than the single-gene traits transferred to date and may require a different approach. Amphiploids are not always perennial, and as we have seen, they are usually genetically unstable and agronomically undesirable. Backcrossing to wheat usually results in a return to the annual habit. Therefore, Anamthawat-Jonsson (1996) proposed backcrossing instead to the perennial parent — in their research, either *Leymus arenarius* or *L. mollis.* The objective then becomes to improve traits such as grain yield and kernel weight in the perennial species.

Anamthawat-Jonsson (1996) listed the traits to be improved by incorporating wheat germplasm into *Leymus* species: perenniality, grain quality, harvestability, threshability, kernel weight, synchronization of maturity, lodging and shattering resistance, meiotic stability, and, of course, grain yield. She has backcrossed partial amphiploids to *L. mollis,* and the progeny were vigorous with long rhizomes. In light of the many problems that have been encountered in transferring perenniality to wheat, the converse approach — using wheat to improve the perennial species — may have considerable merit. Backcrossing to the perennial will almost certainly produce breeding populations with low average grain yields and a high frequency of shattering, requiring the screening of large numbers of genotypes. Logic suggests that perennial allopolyploid species will be more tolerant of added or substituted wheat chromosomes. As we have seen, *L. arenarius* has been used as a grain crop in the past and efforts to improve it using wheat as a donor parent are underway. *Th. ponticum* is easily crossed with wheat but its high ploidy level and lack of diploidization probably would prevent its use in grain production. Wheat might be used as a donor parent for improving *Th. intermedium* or some hexaploid species of *Elymus* with which it can be crossed.

B. Rye

Diploids: Rye (*Secale cereale*) appears to be at least as promising a candidate for perennialization as is wheat. Its chromosomes are homologous with those of its direct perennial ancestor, *S. montanum.* Both species are diploid and cross-pollinated. Rye is very winterhardy, well adapted for grazing, and useful in weed control because of its allelopathic properties. The Soviet perennial-grains programme included a large effort in rye, and they produced some weakly perennial genotypes that were used in limited production. Later, a decades-long effort to breed perennial rye in Germany met with only partial success. Recently, a perennial rye cultivar, 'Perenne', was released in Hungary for grain and forage production.

Despite initial expectations, no perennial rye cultivar has been used in full-scale grain production. Breeders have been stymied by the tendency of plants in *S. cereale*/*S. montanum* populations to be either fertile and annual or highly sterile and perennial. A chain of translocations involving three of rye's seven pairs of chromosomes separates the two species, and gene(s) from *S. montanum* governing perenniality are located on one or more of the translocated chromosomes. Because meiosis in plants heterozygous for one or more translocations produces many inviable gametes with duplications or deficiencies of chromosomal segments, plants in interspecific rye populations fall into one of three categories: homozygous for the *S. cereale* chromosomal arrangement (fertile, annual); heterozygous for one or more of the translocations (highly sterile); or homozygous for the *S. montanum* arrangement (perennial, fertile).

Plants in this last category would seem to answer the breeder's need; however, they are rare, and the large portion of their genomic content derived from *S. montanum* reduces their agronomic desirability and spike fertility. According to Reimann-Philipp (1995), seed-set in *S. montanum* itself is low — approximately 80%. Reimann-Philipp (1995) selected intensely for the *S. cereale* phenotype within an interspecific, perennial population. This population was presumed to be homozygous for the three *S. montanum*-derived chromosomes involved in the translocations, identified as $4R^{mon}$, $6R^{mon}$, and $7R^{mon}$ by Koller and Zeller (1976) but referred to as $2R^{mon}$, $6R^{mon}$, and $7R^{mon}$ by Reimann-Philipp (1995). He was attempting to keep these chromosomes fixed while restoring completely the other four chromosome pairs from *S. cereale* through recombination and selection. But he could not achieve a kernel weight greater than 15 mg (compared with typical values of 40 mg for annual rye under those conditions.)

As an alternative, Reimann-Philipp (1995) proposed selection for perennial plants carrying chromosomes 2R, 6R, and 7R of *S. cereale*. Presumably, such plants would arise from recombination within the ring of six translocated chromosomes. Dierks and Reimann-Philipp (1966) had postulated that perenniality was governed by a single gene that lay approximately 10 crossover units from one of the breakpoints. Selection for perenniality would be routine, but selection for the *S. cereale* chromosomal constitution would require either a laborious testcross procedure or a cytological test. A morphological difference between chromosomes 6R and $6R^{mon}$ did not prove satisfactory for this purpose. Today, the extensive genetic map of rye could allow marker-assisted selection for the *S. cereale* arrangement. A large initial experiment could provide much more detailed information on the genetic control of perenniality; as noted by Reimann-Philipp (1995), the trait is probably affected by more than one gene.

Yet another strategy was followed by L.F. Myers and R.J. Kirchner in the breeding of 'Black Mountain' perennial rye in Australia: backcrossing the interspecific hybrid twice to the *S. montanum* parent. Perennialism (and, presumably, the *S. montanum*-type chromosomal arrangement) was quickly restored by backcrossing. But this cultivar was intended primarily as a forage grass, with *S. cereale* donating genes for nonshattering rachis and improved seed production. Oram (1996) practiced six cycles of half-sib family selection for grain and forage yield in Black Mountain, achieving gains in both traits while maintaining a low level of shattering. With grazing, stands of 'Black Mountain' decline after 3 years; however, if shattering is permitted, stands can be continually replenished by volunteer seedlings. Without selection to develop a cultivar strictly for grain production, we cannot know whether backcrossing to perennial rye while selecting for alleles from annual rye can achieve sufficient yield improvement.

Reimann-Philipp (1995) warned of a hazard when growing diploid perennial rye with the *S. montanum* chromosomal arrangement on a field scale. If pollen from the perennial drifted into seed production fields or breeding nurseries of annual rye, the resulting translocation heterozygosity would seriously and irreversibly degrade fertility in subsequent generations.

Tetraploids: In an effort to improve kernel weight, Reimann-Philipp (1995) used colchicine to double the chromosome number of a perennial *S. cereale/S. montanum* population homozygous for the $4R^{mon}$, $6R^{mon}$, and $7R^{mon}$ chromosomes. The resulting tetraploid, named 'Permontra', had a kernel weight of approximately 30 mg (double that of the diploid), and first-year grain yields over 2000 kg/ha when grown in Germany. Yields declined in subsequent years. 'Permontra' achieved a similar grain yield in the Land Institute's plots in Kansas, but only 15 to 20% of the plants regrew in the next season. After a first-year harvest in the hot, dry summer of 2001, a stand of Permontra at The Land Institute died out completely by September.

Another problem with 'Permontra' — poor seed set — is also common in annual tetraploid rye, because the formation of multivalent chromosomal associations leads to production of gametes with extra or missing chromosomes. Selection can improve seed-set in tetraploid rye, and Reimann-Philipp (1995) pointed out that selection for seed set or meiotic stability can be practiced much more effectively in a perennial, by screening phenotypically or cytologically in one flowering cycle and intercrossing selected plants in the next. He found that phenotypic selection greatly improved seed-set in the perennial spring rye 'Soperta', which was derived from seven 'Permontra' plants that did not require vernalization in order to flower. Another approach would be to introduce the *Ph1* gene from wheat into a tetraploid hybrid between *S. cereale* and *S. montanum* to enforce diploid pairing and improve fertility. The *Ph1* gene was shown to operate when the chromosome carrying it was added to rye.

Tetraploid perennials may provide advantages beyond increased kernel weight. They are reproductively isolated from diploid annual seed production fields. In one study, 'Permontra' had greater heat and drought tolerance, and a much more extensive root system, than did diploid or tetraploid annuals. But perennial rye, whether tetraploid or diploid, will not be grown widely as a grain crop until the problems of sterility, persistence, and maintenance of yield over seasons are solved.

Triticale: To date, the only species to be synthesized by artificial hybridization for use as a cereal crop is triticale (X *Triticosecale*), an amphiploid of durum wheat (*T. turgidum*) and *S. cereale.* (Octoploid triticale cultivars — *T. aestivum/S. cereale*— have also been produced but not widely used commercially.) Triticale has not become one of the world's leading cereals, but its modest success suggests the possibility of developing perennial *T. turgidum/S. montanum* triticales. Derzhavin (1960b) produced and intercrossed many amphiploids derived from crosses between durum wheat and perennial rye accessions. He augmented the gene pool by allowing the amphiploids to pollinate a large number of different wheat/rye F1 hybrids. But the resulting populations were only weakly perennial. Three-way hybrids — from crosses between wheat/*Th. intermedium* hybrids and perennial rye— were more strongly perennial but had sterility, low yields, and small seeds.

Robert Metzger (USDA-ARS retired, Corvallis, OR, personal communication) reports that a *S. montanum*-derived triticale that he has developed is not sufficiently perennial, but he recommends producing and screening more new amphiploids involving a wider range of *S. montanum* germplasm. It could be that no triticale will be fully perennial, having only one of three genomes derived from a perennial species. Intercrossing of diverse lines followed by selection could improve persistence over years. If improved hexaploid or tetraploid perennial wheats can be developed, they could be crossed with *S. montanum* to produce more strongly perennial triticales.

In attempting to develop perennial triticales, breeders can take lessons from development of the annual crop. Primary triticales, *i.e.*, newly doubled wheat/rye hybrids, inevitably suffer from sterility, seed shrivelling, lodging, and low yield potential. Decades of intense selection and introgression have resulted in triticales that are cytologically stable and improved for all of these traits, but they stand on a narrow germplasm base. Improving the performance and genetic variability of the triticale gene pool can be accomplished by several means: production of new primary triticales; triticale/wheat crosses; triticale/rye crosses; and crosses between hexaploid and tetraploid triticales.

Primary triticales derived from the wild *S. montanum* are even more agronomically primitive than *S. cereale*-based primary triticales and will require even greater breeding effort with a wide range of parents. "Substituted" triticales in which one or more wheat chromosomes replace those of rye are often agronomically superior; chromosomes 2D and 6D appear to be selectively propagated by breeders in populations segregating for R- and D-genome chromosomes. But it must be kept in mind that until genes conditioning perenniality can be mapped, random substitution of wheat for *S. montanum* chromosomes will reduce the chances of selecting a strongly perennial triticale.

Direct Domestication of the Perennial Triticeae

Intermediate wheatgrass: From domestication of perennial grasses with wheat or rye as a donor parent, it is a relatively short leap to domestication without any interspecific crossing. Three large-seeded perennial species that have been hybridized with wheat also have attracted attention as candidates for direct domestication. By far, the most work in this area has been done with *Th. intermedium* by Wagoner (1990a, 1995) at the Rodale Institute in Pennsylvania and her colleagues at the USDA-NRCS Big Flats Plant Materials Centre in New York. Wagoner (1990a) described in detail the characteristics that make intermediate wheatgrass a good candidate for domestication as a perennial grain, while noting shortcomings that must be addressed. Becker *et al.* (1991) concluded that its grain has protein quality "superior to the cereal grains now commonly grown", with no significant amounts of antinutrients. Recurrent selection is a logical breeding method for improving an only slightly domesticated, cross-pollinated species like *Th. intermedium.* Using mass selection without controlled pollination, Knowles (1977) increased seed yield in an intermediate wheatgrass population by 10% per cycle. In each cycle, 1000 plants were evaluated for spike fertility over three years, and the best 50 were selected. When selected plants were removed to the greenhouse over the winter to exclude pollination by non-selected plants, thereby doubling parental control, gain per cycle increased to 20% — a result perfectly consistent with selection theory. Gridded mass selection is used to exercise control over microenvironmental effects and increase selection response. Wagoner (1990a, 1995) evaluated 300 accessions of intermediate wheatgrass, for grain yield, yield components, and end-use quality, selecting the 20 best accessions in 1989. The selections were transplanted into a polycross nursery, and 380 progeny resulting from pollination among the selections were evaluated, in a field divided into blocks of 25 plants each, between 1991 and 1994.

The best 11 plants resulting from within-and among-block selection, plus three selections resulting from further evaluation of other accessions, were put into a second-cycle polycross, and 400 individual progenies were evaluated in a second blocked nursery. Yield per plant in the 14 selections was approximately 25% higher than the population mean. Evaluation of the second-cycle population is underway, and selected plants will be intermated in 2002 to complete another breeding cycle (M. van der Grinten, USDA-NRCS, Big Flats, NY, pers. commun.) Better environmental control through selection within blocks may have produced the five-percentage-point improvement in selection response over that of Knowles (1977), but it remains to be seen if the small effective population size in the second cycle (14 plants) will restrict genetic gain in the future.

Wildrye: The Land Institute in Kansas has studied perennial cool-season grasses as potential grain crops for over 20 years. They evaluated almost 1500 accessions representing 85 species of *Agropyron, Thinopyrum, Elymus,* and

Leymus, along with 2630 accessions of other species, between 1979 and 1987. The species selected as having the greatest potential for domestication was *L. racemosus*, known commonly as giant or mammoth wildrye. However, prospects for utilization of this species in the near future are unclear. Among 16 accessions evaluated over 2 years, yields did not exceed 830 kg/ha, and yield declined rapidly in the second and third years. Wildrye's great vigour, accompanied by large spikes but sparse seed-set, resulting in low harvest index, may provide considerable scope for breeders to select for diversion of photosynthate towards grain production. But there is no current breeding programme for grain yield in *L. racemosus;* until selection is undertaken, no conclusions can be drawn regarding its potential. *L. racemosus* is self-pollinated and would require a breeding approach different from that taken with *Th. intermedium.*

Lyme grass: Lyme grass or beach wildrye (*Leymus arenarius*) has been used as a food grain since the time of the Vikings, and, as we have seen, is being studied as a potential grain crop in Iceland. There is significant genetic variation among accessions of *L. arenarius* and *L. mollis,* and it would be interesting to know which approach would result in more rapid genetic progress: direct selection within the species or an interspecific backcross programme using wheat as a donor parent. The latter strategy takes advantage of genes selected through millenia of wheat domestication and breeding, but introduces chromosomal instability.

Prospects for direct domestication: Is there sufficient genetic variation within these three, or other, cool-season grasses to support large improvements in yield, kernel weight, and other traits? The very existence of annual grain crops proves that selection over thousands of years can move the mean of a species far beyond its original phenotypic range. Gains of 20 to 25% per cycle are much more rapid than typical gains in major annual crops, even considering the longer selection cycle of perennials. But to effect sufficient changes in a matter of decades rather than centuries — while possibly working "uphill" against the problem of resource allocation in perennials — will require much larger breeding efforts than have been undertaken to date. Relatively small efforts at domestication, which are within the capabilities of nonprofit organizations such as the Rodale Institute or Land Institute (or small-scale breeding programmes within larger organizations such as USDA or universities), must be expanded to a much larger scale by university, government, or corporate breeding programmes if wholly new perennial grain crops are to be developed.

The yield increases of 25% per cycle achieved by Wagoner *et al.* (1996) are remarkable, especially considering that they selected for other traits in addition to yield. But because response to recurrent selection tends either to follow a linear path or decelerate, future gains per cycle will probably be no greater than a constant percentage of the base population's yield. If a hypothetical perennial grass population yielding 500 kg/ha of grain undergoes selection, with a yield increase of 125 kg/ha/cycle (25% of the base yield), 20 cycles will be required to reach 2500 kg/ha. Because selection in perennials must be based on evaluation over two or more seasons, a single cycle can occupy four or five years. Obviously, if it is going to take almost a century to develop a high-yielding perennial crop through direct selection, a long-term commitment is required; however, such a rate of progress is much greater than the rate at which our annual crops were domesticated and improved. Marker-assisted selection and/or some genetic input from wheat could speed up the process.

Oat: Perennial oats for grain production might be developed from crosses between the cultivated hexaploid oat (*Avena sativa,* 2n=42, genomes AACCDD) and a wild, perennial, autotetraploid relative, *A. macrostachya* (2n=28, CCCC). Such crosses require embryo rescue.

The F1 is highly sterile, but backcrosses to *A. sativa* have been made, and limited pairing between chromosomes from different parents does occur. J. P. Murphy has produced a 70-chromosome amphiploid between the species. The objective of this cross is to improve winterhardiness in the annual crop; perenniality has not been evaluated. Because of the partial homology that exists between chromosomes of the parental species, the amphiploid is likely to suffer from the same chromosomal instability found in wheat amphiploids. But the amphiploid, like the hybrid, can be backcrossed to *A. sativa* (Murphy, pers, commun.) Ladizinsky (1995) domesticated accessions of two wild annual oat species by using the cultivated oat A. sativa as the donor of genes for nonshattering and other traits. Perhaps this approach could be tried with *A. macrostachya.*

More hybrid combinations and larger populations will be needed if genetic studies and selection for perenniality are to succeed in the backcross generations. A more diverse sample of *A. macrostachya* parents would be desirable, but the species is restricted to two mountain ranges in Algeria, limited germplasm collections exist in the United States, and there are very few accessions held in other countries. Selection for perenniality in colder climates could be thwarted by the lack of winterhardiness in oats; winter annual oats are not generally sown above 35 degrees latitude in North America. At sites where *A. macrostachya* was collected by Guarino *et al.* (1991), the mean minimum temperature of the coldest month ranged from-0.6 to-3.6°C.

Rice: Although tropically adapted, rice (*Oryza sativa*) has the C3 carbon fixation pathway and is included here with the cool-season grasses. The perennial ancestor of *O. sativa* is *O. rufipogon.* Both species are diploid (2n=24) with homologous chromosomes and they can be hybridized easily. Indeed, natural hybridization and introgression occur in the field. From 1995 to 2001, the International Rice Research Institute (IRRI) had a programme for development of perennial rice cultivars to reduce erosion on the steep slopes where upland rice is often grown. Populations from IRRI's breeding programme, which was discontinued in 2001, have been distributed to cooperators in China, where perennial rice breeding efforts continue.

Sacks *et al.* (2000) found wide variation in second-year survival among 51 *O. sativa/O. rufipogon* F1 hybrids. Sixteen percent of the hybrid combinations had greater than 50% survival, and 19% of all hybrid plants survived. In a cross between a rice cultivar with a regeneration score of 1.0 and an accession of *O. rufipogon* with a score of 3.8, the F1 had a regeneration score of 4.0, and the scores of F2 clones ranged from 0 to 5. Paradoxically, in three of the four chromosomal segments that affected regeneration ability, it was the annual parent's allele that had a positive effect. In contrast to mapped regrowth loci in *Sorghum bicolor/S. propinquum* populations, none of the regeneration loci were associated with effects on tiller number.

Selection in interspecific populations may be aided by rice's detailed molecular map and the known locations of chromosomal segments affecting traits of domestication.

Many of the traits separating the annual and perennial species show polygenic inheritance. Surprisingly, *O. rufipogon* was the source of four chromosomal segments with positive effects on testcross grain yield in one set of backcrosses; however, three of the four segments were adjacent to segments that either increased plant height or delayed maturity. With positive alleles affecting perenniality *and* productivity apparently being contributed by *both* parental species, prospects for breeding high-yielding perennial rice genotypes may be bright.

Oryza sativa can also be hybridized with *O. longistaminata,* the perennial ancestor of West African rice, *O. glaberrima.* The perenniality of *O. rufipogon* lies in its ability to regrow repeatedly through production of new tillers, whereas *O. longistaminata* regrows from rhizomes. In crosses between *O. sativa* and *O. longistaminata,* genes affecting rhizome production appear to be linked to genes for hybrid embryo abortion. Consequently, IRRI scientists backcrossed rare hybrids to both parental species and intercrossing the progeny in an effort to develop a rhizomatous, agronomically acceptable genotype. Tao *et al.* (2001) recovered a single rhizomatous individual from among 162 plants produced by backcrossing an *O. sativa/O. longistaminata* hybrid to *O. sativa.*

A long-lived, three-species hybrid (*O. sativa/O. rufipogon/O. longistaminata*) has persisted through winters in China with monthly mean temperatures as low as 5°C. Crossing the hybrid with *O. sativa* and intermating perennial progenies has eliminated shattering. One possibility for breeding an even more cold-tolerant perennial rice exists. An ecotype of *O. rufipogon* known as 'Dongxiang' has the ability to regrow in regions of China where temperatures below -10°C are common.

BREEDING PERENNIAL GRAINS: WARM-SEASON GRASSES

SORGHUM

Hybridization with Sorghum propinquum: In tropical environments, grain sorghum (*S. bicolor,* 2n=20) is able to regrow from basal nodes to produce a rattoon crop. But breeding a sorghum that is winterhardy in temperate regions will require transfer of genes from related species.

A perennial native of southeast Asia, *S. propinquum* is rhizomatous and diploid, with chromosomes largely homologous to those of grain sorghum. Paterson *et al.* (1995) evaluated rhizome-related traits of 370 F2 and 378 BC1 plants from a cross between the two species. Surviving a mild winter in southern Texas, USA, with only three nights reaching temperatures of -3°C to -4°C, 92% of F2 plants and 46% of BC1 regrew in the spring. Plants regrew either from tillers or from rhizomes. Forty-eight F2 plants representing the range of the population were selected for progeny testing.

From all F2 plants and F3 lines, Paterson *et al.* (1995) collected data on number of rhizomes producing above-ground shoots, distance between the centre of the crown and the most distal shoot, a subterranean rhizome score, tillering, and regrowth. They mapped chromosomal segments affecting these traits in the F2 plants and F3 lines, using 78 RFLP loci.

Rhizomatousness was a complex trait, with nine different chromosomal regions on seven of sorghum's ten chromosomes having detectable effects on at least one of the rhizome traits. Individual segments accounted for between 5 and 13% of the total variation. All but one of the seven segments associated with regrowth was also associated with one or more rhizome traits, and all four segments associated with tillering were also associated with regrowth or rhizomatousness.

Because *S. propinquum* is a tropical species, rhizomatous progeny of crosses between *S. bicolor* and *S. propinquum* would probably not be winterhardy at middle or northern latitudes, without successful selection for deeper rhizome growth. But there is great potential for developing a perennial grain sorghum for the tropics or subtropics from such populations.

Hybridization with johnsongrass: Johnsongrass (*S. halapense,* 2n=40) is a tetraploid, probably an amphiploid that combines the genomes of *S. bicolor* and *S. propinquum*. It is a very strong and aggressive perennial, and a notorious weed. Like *S. propinquum,* johnsongrass stores starch in its rhizomes. As a consequence, its rhizomes have no cold hardiness, unlike those of temperate grasses, which store fructosans. Natural selection for deeper-growing rhizomes has allowed johnsongrass to spread as a weed as far north as Ontario. Although the most northerly biotype reproduces mainly by seed, regrowth from rhizomes occurs throughout the range of the species.

Early research on hybrids between diploid sorghum and johnsongrass produced two types of hybrids: 30-chromosome plants that were male-sterile but could be backcrossed to the diploid parent and fertile 40-chromosome plants derived from unreduced female gametes in the diploid parent. The 30-chromosome plants were more strongly rhizomatous. Hadley and Mahan (1956) identified seven 20-chromosome backcross plants that were rhizomatous, but most were chlorophyll mutants. Three years of selection failed to produce a single diploid line that was rhizomatous.

Hybrids between *S. halapense* and induced tetraploid lines of *S. bicolor* are easily made. In an effort to produce a perennial grain sorghum at The Land Institute, Piper and Kulakow (1994) crossed *S. halapense* with tetraploid grain sorghum lines. In an interspecific F3 population, approximately 40% of plants were rhizomatous. There was no significant negative correlation between rhizome production and grain yield in the F3 generation, but a negative association arose with backcrossing. Yield was strongly related to plant biomass and root biomass, both with pheontypic correlations of 0.70. The F3 population — selected for winterhardiness but not for yield — had a grain yield 62% as high as the mean non-irrigated sorghum yield in Saline County, Kansas, where the experiments were conducted. But rhizome production dropped to near zero in other populations derived by backcrossing to tetraploid *S. bicolor* in an effort to increase grain yield.

Land Institute breeders have selected for winter survival among the rare rhizomatous BC2 plants and their selfed progeny. The phenotypes of winterhardy selections remain very distant from that of the cultivated parent, despite the latter's expected 87.5% genetic contribution to the BC2. The selections are taller and later maturing than either parent, have open panicles and small seed, and produce many tillers, although not as many as johnsongrass. Rhizome mass is less than 10% that of johnsongrass — sufficient for overwintering, but not enough to allow interspecific progenies to become aggressive weeds.

Both rhizomes and tillers originate from meristems at the base of the plant, and there appears to be considerable overlap in their genetic control. It may not be possible, or even desirable, to select a low-tillering, sufficiently rhizomatous genotype. Indeed, selection for yield improvement may be more effective in highly tillering populations. If a grass plant is regarded as a population of largely autotrophic tillers, and each tiller supports both seeds and rhizomes, then the most direct route to increased yield is via additional tillers. Of course, this implies increased biomass.

Piper and Kulakow (1994) concluded that development of a winterhardy sorghum (*i.e.*, one that produces 80g of rhizomes per plant) with a grain yield of over 4000 kg/ha is feasible, through selection for greater biomass and reallocation of photosynthate to seed production. The foundation germplasm for breeding a perennial sorghum may necessarily consist of high-biomass plants that produce more tillers than annual sorghum. With a sufficiently large genetic base, subsequent selection for improved harvest index and seed size could be successful. Because there is some homology between the chromosomes of grain sorghum and johnsongrass, multivalent chromosome associations are common at meiosis in interspecific tetraploids. Multivalents, in turn, cause poor seed set because of nondisjunction of chromosomes. Luo *et al.* (1992) demonstrated that selection for fertility can be effective in autotetraploid grain sorghum, which is generally plagued by low seed set. Breeders could cross perennials with the highly fertile tetraploid germplasm that Luo *et al.* (1992) have produced. Broadening and improving the genetic base of tetraploid perennial sorghum will require introduction of more agronomically elite germplasm. One rapid method of incorporation would be to pollinate both diploid and induced-tetraploid strains of elite, large-seeded inbred lines with the best tetraploid perennials. From the diploid/tetraploid crosses, breeders can select 40-chromosome hybrids that arise from unreduced gametes. The Land Institute is now taking this approach to develop genetically diverse breeding populations.

Pearl Millet

Pearl millet (*Pennisetum glaucum,* 2n=14), like sorghum, is a tropical, annual diploid with a perennial, tetraploid relative. Napiergrass, *P. purpureum* (2n=28) has one genome homologous and one nonhomologous to that of pearl millet. Amphiploids resulting from colchicine treatment of hybrids between the species are male and female fertile. Hanna (1990) backcrossed these hexaploids to diploid and tetraploid pearl millet lines and produced perennial progeny; however, both types of backcross plants (tetraploid and pentaploid) were highly sterile and unable to survive the mild winters of south Georgia, USA. Napiergrass, the only known species in pearl millet's secondary gene pool, is not rhizomatous, so selection for winterhardiness would probably not be successful. A perennial millet for grain production in the tropics is a reasonable prospect.

Dujardin and Hanna (1990) interpollinated hybrids and their derivatives from crosses between tetraploid pearl millet and *P. squamulatum* (2n=54), a more distant, perennial, apomictic relative. Some progenies (2n=48) were both perennial and apomictic. Apomixis can be used to ensure grain production and genetic stability in highly heterozygous progenies of interspecific crosses. *Pennisetum squamulatum* is also non-rhizomatous, and its progeny are not likely to be perennial outside of the tropics.

Maize

Hybridization with tetraploid perennial teosinte: Efforts to develop perennial maize (*Zea mays* ssp. *mays,* 2n=20) have been sporadic at best; as in other crops, hybridization between maize and perennial relatives has led primarily to improvement of the annual crop. Shaver (1964) first attempted development of maize-like perennials from crosses between colchicine-induced tetraploids of maize and a wild, perennial, tetraploid relative, *Z. mays* ssp. *perennis* (2n=40). Selection within the resulting tetraploid populations and backcrosses to tetraploid maize effectively increased the frequency of perennial progeny. Crosses to diploid maize produced perennial triploids, but all diploid selections were annual. Shaver (1967) combined a postulated gene (*pe*) for perenniality with recessive genes for indeterminacy (*id*) and grassy tillers (*gt*) in a diploid background, to produce perennial plants; however, the *idid* genotype prevented production of ears. Because Shaver (1967) had developed a separate *idid* population in a different genetic background that did produce ears, he suggested that perennial diploids could also be made fertile if the genetic background were manipulated.

Hybridization with diploid perennial teosinte: Little further attention was paid to perennial maize until the dramatic discovery of a diploid species of perennial teosinte, *Z. mays* ssp. *diploperennis*. Initial studies showed that inheritance of perenniality was relatively simple in maize/*diploperennis* crosses, but perenniality was inferred from tillering habit, a potentially misleading technique. In subsequent, larger-scale experiments, inheritance of tillering in progeny of similar inter-subspecific crosses was more complex, and perennial maize types were not recovered even in large segregating populations.

Genetic mapping in maize/annual teosinte crosses show that most traits of domestication separating the species are oligogenic, and the loci tend to be clustered on the map, through either linkage or pleiotropy. A similar study of these traits, plus perenniality, in crosses between maize and diploid perennial teosinte would be of great value to any breeding programme attempting to combine perenniality with the agronomic phenotype of maize. This would require substantial effort to evaluate large segregating populations for tillering, rhizome production, and capacity to produce seed over multiple seasons. Once the genomic regions of interest are identified, marker-assisted selection can be used to incorporate them into a maize background and eliminate unwanted alleles such as those conditioning hard glumes and shattering.

One serious obstacle to adoption of any teosinte-derived perennial grains is the lack of winterhardiness of these tropical species. There are no winterhardy species of *Zea*. Because the bulk of maize production and breeding occurs in temperate areas, there has been little incentive to develop perennials from crosses with *Z. mays* ssp. *diploperennis*.

One possible approach has not been suggested to date: selection for rhizome depth. As we have seen, johnsongrass rhizomes also are not winterhardy if near the soil surface, but dispersal of the species into higher latitudes has been made possible by selection for deeper rhizomes. Superimposing selection for this undoubtedly complex trait on selection for perenniality and traits of domestication, not to mention yield, may entail a much larger effort than any breeding programme is willing to undertake.

Hybridization with eastern gamagrass: The closest winterhardy relatives of maize are in the genus *Tripsacum*. Eastern gamagrass (*T. dactyloides*), for example, is currently grown as a perennial forage grass as far north in the western hemisphere as Kansas and Massachusetts, and can be grown in the Corn Belt. *T. dactyloides* has been hybridized many times with maize, beginning with the work of Manglesdorf and Reeves (1931). Plants of the diploid (2n=36) or tetraploid (2n=72) races may be crossed with maize. If *Tripsacum* is used to pollinate maize, embryo rescue is necessary, but if maize is used as the male, some hybrid seed may be obtained without rescue. In addition, several strains of popcorn, when pollinated with tetraploid *T. dactyloides,* produce large amounts of hybrid seed that does not require embryo rescue. Some have good crossability with diploid *T. dactyloides* as well. Contrary to typical results, Eubanks (1995, 1997) reported that a putative 20-chromosome hybrid between *T. dactyloides* and *Z. diploperennis* showed 93 to 98% pollen fertility.

Natural introgression between *Tripsacum* and maize has not been observed, but morphological and molecular evidence supports the hypothesis that the species *T. andersonni* is an intergeneric hybrid containing three genomes (54 chromosomes) from *Tripsacum* and 10 chromosomes from *Zea* in *Tripsacum* cytoplasm. The uniformity of this ancient natural hybrid indicates that *T. andersonii* arose from a single hybridization. It has been able to spread across tropical Latin America because of its vigorous perenniality.

In addition to being perennial, tetraploid *T. dactyloides* is a facultative apomict. Perennial hybrids result from artificial crosses between tetraploid *Tripsacum* and maize, and some seed-set can result from apomixis. The hybrids, derived from parents with different basic chromosome numbers and chromosomes of different sizes (those of maize being larger), are male sterile, with cytological behaviour that is anything but regular. Harlan and deWet (1977) summarized methods for utilizing such hybrids in maize improvement. Either 28-chromosome or 46-chromosome hybrids — derived from diploid and tetraploid *T. dactyloides* parents, respectively — can be backcrossed to maize. In either case, *Tripsacum* chromosomes are eliminated with backcrossing. Elimination occurs more gradually in progeny of 46-chromosome hybrids, and the 20 chromosomes of the resulting backcross plants can contain significant genetic material from *Tripsacum*. All 20-chromosome backcross plants derived to date have been annual and non-apomictic. Kindiger *et al.* (1996) derived an annual, 39-chromosome line that carried 9 *Tripsacum* chromosomes and displayed an intermediate level of apomixis. Most hybridization with *Tripsacum* has been for the purpose of either elucidating the evolution of maize or transferring resistance or other genes to annual maize. The latter purpose implies backcrossing to maize. But development of perennial populations may require interpollinating plants in early backcross generations that still carry many *Tripsacum* chromosomes, or even backcrossing to *Tripsacum*. As Harlan and deWet (1977) commented,

"Apparently, if one wishes to contaminate maize with *Tripsacum* one should first contaminate *Tripsacum* with maize."

New approaches to perennial maize are being explored. An anomalous fertile hybrid between diploid *T. dactyloides* and maize was discovered by one of the authors (BEZ) in 1997 near the mouth of the Big Nemaha river in Richardson County, Nebraska, USA. This derivative of natural introgression between *T. dactyloides* and a putative commercial hybrid is being hybridized with gynomonoecious *Tripsacum* — both diploid and tetraploid — and with tassel-seed popcorn to develop a 56-chromosome perennial cultivar for production of grain, forage, fibre, and fuel. The difficulties encountered in introgressing apomixis from *Tripsacum* into maize should temper hopes for a rapid synthesis of perenniality with high grain yield. Whatever the initial population, and even with marker-assisted selection, the process of recovering perennial, winterhardy segregants with maize-like ears, and then breeding for yield and other agronomic traits will be long and arduous.

Direct Domestication of Warm-season Grasses

Eastern gamagrass: Could *T. dactyloides* be domesticated directly, without introgression of genes from maize? To do so would be an accomplishment parallel to that of domesticating maize from annual teosinte — a feat requiring thousands of years and producing genetic and physiological changes much greater than those involved in domestication of Asian cereals such as wheat and rice. To develop a crop from eastern gamagrass using the knowledge and techniques provided by 21st-century genetics, while leaving aside the important genes of domestication available in maize, would be an ambitious project.

Wagoner (1990a) described in detail the status of eastern gamagrass as a potential grain crop, and the species' most discouraging characteristic: very low seed yield. Interest had been stimulated by the discovery of a gynomonoecious, or pistillate, mutant in which pistillate and perfect spikelets replace the staminate spikelets of the normal inflorescence. The result is an increase of up to 20-fold in the number of seeds produced per plant; however, the seeds are small, so that the weight of seed produced per plant is increased by only a factor of 3.

Plants of eastern gamagrass are large, vigorous and widely adapted. The increase in sink size made possible by the pistillate mutant may provide an opportunity to increase seed yield dramatically through increased harvest index— the yield component usually found to have had the greatest effect on yield improvement in traditional grain crops.

Furthermore, Jackson and Dewald (1994) found that the increased seed yield of pistillate plants did not come at the expense of plant vigour or longevity. Carbohydrate reserves were significantly higher in pistillate than in normal genotypes. For breeders, there is a huge pool of genetic variability available in the species. Although tetraploid *T. dactyloides* reproduces apomictically, parental combinations can be produced via BIII hybrids, in which an unreduced egg is fertilized by a haploid sperm. Alternatively, obligately sexual tetraploid plants can be produced via colchcine treatment of diploid *T. dactyloides* plants, all of which are sexual. The genus *Tripsacum* contains many species that lack winterhardiness but have desirable traits that, potentially, could be transferred to eastern gamagrass: synchronous flowering, large spikelet number, higher seed yield, and other variations in plant morphology.

The maximum seed yield of eastern gamagrass in plots at The Land Institute has been 240 kg/ha, in the third year after sowing. It remains to be seen how rapidly yield can be improved through selection within pistillate populations. And there is another question: would the 10-, 20-, or 30-fold yield improvements required to make eastern gamagrass a viable grain crop have larger negative effects on plant vigour and persistence than did the three-fold yield boost brought about by the pistillate mutation? That increase was large relative to the grain yield of a normal plant but required diversion of only a small amount of photosynthate, relative to the plant's large biomass.

The food quality of Eastern gamagrass is excellent. But, even if yield can be improved, other problems must be solved. One problem is disease. Infection by maize dwarf mosaic virus B has been very serious in plots at The Land Institute. Also, the hard fruitcase of *Tripsacum* weighs almost three times as much as the seed itself and makes processing difficult. It was an extremely rare mutation in annual teosinte that freed the kernel from the fruitcase and allowed its use as a grain and the development of maize. This may be a gene that breeders will be forced to transfer from maize.

Indian ricegrass: Grain is currently being harvested from a perennial grass for human food in northeastern Montana, USA. Indian ricegrass (*Oryzopsis hymenoides*), cultivar Rimrock, has reduced seed-shattering and produces gluten-free grain.

The grain is being produced, milled, and marketed under the trade name Montina. Yields vary between 250 and 500 kg/ha, but improvement through breeding may be feasible. Germplasm collections exhibit great phenotypic diversity, and very large-seeded genotypes are known (T.A. Jones, USDA-ARS, Logan Utah, pers. commun.). Certainly, other perennial grasses native to the western USA could be considered for domestication as grain producers, but no attempts have been made.

TECHNOLOGY: TOWARDS AN EVOLUTIONARY PLANT BREEDING

Commercial varieties enter the market through a system of trials and official release. That system requires uniform and stable varieties, and the breeding work must be streamlined for such end results. But local selection is not constrained by those requirements and therefore enjoys some freedom that does not exist within the formal system. Farmer-breeders are free to distribute heterogeneous varieties and can allow crops to continue evolving.

Exploiting heterogeneity and crop evolution in farmers' fields are outside the scope of most plant breeding research. One exceptional experiment, however, has shed some scientific light on the issue. It was started at the University of California (UC) in 1928. Composite cross populations of barley were produced, some of which were extremely diverse in origin of sources. These populations were exposed to continuous natural selection in current modern farming environments and became the subject of studies during the career span of several generations of UC professors. It appears that after low yields in initial years, the composite cross populations gradually improved in performance and eventually became quite good yielders, with excellent yield stability and disease resistance. These results inspired Suneson (1956) to propose an evolutionary plant breeding method. After assessment of later generations of the same material, Soliman and Allard (1991) concluded that such evolutionary breeding "is unwarranted" if yield potential is the major goal. However, if disease resistance and yield stability are two major objectives, "the composite cross approach is an efficient method". This amounts to saying that a major part of world agriculture, many high-input systems included, could be well served by this approach.

In short, this means constructing a body of broadly diversified germplasm and exposing it to natural selection in areas of contemplated use. For those who are familiar with traditional farmers' breeding, this may sound like reinventing the wheel. In fact it is an improvement of the old wheel of plant breeding. The first step, constructing a body of of broadly diversified germplasm, is not all that straightforward. Science has access to world collections and information sources that are unavailable to farmers. A research institute can chose relevant germplasm and make composite cross populations with an evolutionary potential, most probably far beyond that of locally available varieties.

The immediate outcome, the early generation composite cross population, will be unadapted everywhere and is likely to yield poorly. With time, however, recombinations and natural sorting will improve the adaptation, and, according to the Californian experience, narrow the gap with commercial varieties. The long term outcome could be populations that outperform commercial varieties in disease resistance and yield stability and that may be used as a source of artificial selection for high yield.

The disease resistance appears to have evolved through the building up of polygenic complexes. Therefore, it provides a durable resistance as opposed to the monogenic and, therefore, mostly non-durable resistance usually bred into commercial pure line varieties. The stability, at least to some degree, depends on the buffering effect of crop heterogeneity. The Californian experiment shows that when the population is propagated in isolation for a very long time, diversity will start declining, resulting eventually in reduced stability.

Taking the lessons from these experimental findings into the context of current development needs, a few conclusions can be drawn. We need breeding populations with a very high evolutionary potential, and these populations must be exposed to the stress conditions of, or similar to, current farm environments. Furthermore, a certain level of diversity within populations must be maintained in order to sustain evolutionary potential and yield stability. Landraces are usually found to be heterogeneous. But do they have the evolutionary potential required by current development needs? Are they being subjected to a selection pressure that ensures yield stability, and are they as high yielding as they could be?

Scientific evidence may not be available to give direct answers to such questions. Community visits may not be helpful either. A confusing picture of different selection practices and frequent change of seeds (and therefore lack of persistent long term selection), appears in many communities where traditional seed systems prevail. It is also common to see the coexistence of modern and traditional varieties. More systematic efforts and a certain level of organization are necessary to make full use of knowledge already existing within a culture, to exploit fully the potential within the locally available germplasm, and to take advantage of opportunities provided by science.

Community Organization

Widely diverse forms of organizations dealing with seed management have sprouted up at the community level in recent years. I will present an Ethiopian case representing a traditional society, and a Philippino case representing a modern society. Criteria of classifying these societies as traditional and modern are access to external markets and farm inputs. These were absent in the Ethiopian case and present in the Philippino context.

The Ethiopian case is from Tigray, in the middle of the famous Abyssinian gene centre. Renowned for genetic richness and for knowledge and culture related to seed management, the area should be expected to provide an excellent site for community seed banks. But seed banks were organized more because of poor seeds than because of genetic wealth. Poor seeds were seen as one of the reasons for poor agricultural performance. This was in the 1980s and the area was isolated by war. No external support was possible and community leaders had to look for sources of improvement within their own communities. They knew that those sources existed as, in all societies, there were experts known for their skills in traditional seed selection who had fine local seeds. But they needed an organization to extend the benefits of those experts and those seeds to the wider community.

The Philippine case is from Mindanao, from a community where all farmers have formal school education, and where most of them have taken over their farms after the introduction of modern farming. These farmers had no memory of a pre-Green Revolution practices, such as seed selection and traditional seed management. In recent years, however, some of them have switched to organic farming because of declining profit margins in the high-input system. This has created a need for a different type of seed and also an organization to recover and reintroduce the lost traditional seeds as well as to re-establish a local seed system.

These two cases, one isolated from, and the other influenced by, the modern system required different organizational approaches to the seed management problems. In both cases, however, local seeds and on-farm selection were established as the platform on which to build.

In Ethiopia, the felt problem was poverty and recurrent famines. Community leaders saw poor crop performance associated with poor seeds as one of the causes, but also knew about individuals who had good seeds and had a reputation of being excellent seed selectors. They decided to extend the good seeds through a credit scheme. Community seed banks were established and the local experts on traditional seed selection were used to identify good seeds for lending. Unlike genebanks, they were not concerned with conservation, but rather with the circulation of seeds. Like a commercial bank, they put their capital to work. They also had social concerns and gave priority to loan applicants who were poor and had a particular need for access to good seeds (and who needed protection against private moneylenders). The seed banks were owned by the community and controlled by democratically-elected community assemblies.

The first of these seed banks was established in 1988 and, within a few years, was replicated in most local districts of a region of close to four million people. The growth of the community seed banks continued after peace, in 1991. But now the challenge is twofold: the seed banks must develop in order to remain relevant; and they must protect their integrity and independence when government institutions and seed companies start appearing in the area.

During the war, people had no choice. They had to depend on their own resources. And they proved to themselves that the necessary skills and the required good seeds existed within their own community and could be used to solve their immediate problems. Currently, these Ethiopians do not seem to consider their seed effort as an emergency measure that can be phased out when government services and seed companies begin to function and, perhaps, to take over. They want to keep their seed banks as permanent institutions. Their challenge now is to realize the development potential of these institutions and the evolutionary potential of their seeds within the context of an opened-up economy. But the seed banks and their associated seed supply system are also a challenge for the scientific plant breeding system which is now being re-established in the area. That challenge need not render the local seed supply system obsolete by supplying better commercial seeds; scientists could opt to work with local farmer-breeders in order to ensure that seeds offered through the seed banks remain competitive.

The case from the Philippines, the Community Based Native Seed Research Centre (CONSERVE) in Mindanao, arose as a response to critical economic problems of the high-input system of rice cultivation. With increasing prices for inputs and decreasing prices for the produce, profit margins were shrinking, and farmers became dangerously dependent on moneylenders. Some individuals saw a switch to organic farming as the only way out. In that situation, farmers needed a new organization. Unlike the Ethiopian case, where community assemblies representing the entire community were the organizers and owners of the seed activities, this Philippine group was a minority, and membership was individual.

Starting in 1992, this group organized a search for local traditional seeds, which were recovered from farms in isolated remote areas. These seeds were multiplied and distributed to the members for on-farm evaluation and screening. After only two years of operation, I visited the project and found farmers discussing seeds with excitement. From an initial challenge of sorting and selecting among a great number of landraces offered to them, some had already started selecting within landraces, and some had started crossing varieties and keeping written records of what they were doing. They involved their wives and children. More than a change of seeds had occurred; there was a change of mind also. Before, these farmers grew modern varieties. Such varieties are supposed to be pure, and off-types are considered as impurities. In case they were saving seeds, they had been taught to rogue the off-types before harvest to maintain varietal integrity. The change of mind involved seeing diversity in the field as a resource, rather than an impurity, and seeing themselves as active selectors, rather than passive receivers of ready-made varieties.

It is a mistake to consider this as a rejection of science. It is a withdrawal from the commercial seed system. If it is true that scientific plant breeders are working for farmers and not for seed companies, they might find organized farmer groups another outlet for their scientific achievements. That would require the development of participatory plant breeding methods.

8

Genetic Engineering in Plant

GENETIC ENGINEERING AND PLANT BREEDING

Plant breeding is certainly not a new concept. Humans have been breeding and selecting improved plants for centuries, choosing crops with better yield, bigger fruit, or less susceptibility to pests and diseases. This is particularly true of annual crops such as wheat and corn, where many of the cultivars now grown bear little resemblance to those of even last century.

However, progress in the improvement of long-lived tree crops by breeding has been much slower than for annual crops, despite the wide range of characteristics available. The main problem is the length of time from the germination of seeds until the production of fruit. Promising selections then need to be tested on rootstocks, in different locations. The whole process from start to finish may take as long as 15 years. The major challenge presently facing breeders is to shorten the length of time required to develop new cultivars. This is where the use of biotechnology can assist. The objectives of plant breeding and improvement are basically the same, whether they are achieved by conventional breeding methods or genetic engineering. The difference between the two lies in the speed and accuracy with which new cultivars can be achieved. If an established plant cultivar needs a disease-resistance gene, the genetic engineer will be able in the future to simply splice it in without disturbing the 100,000-odd genes already in the plant. However, for the conventional plant breeder, the gene for the characteristic needed comes together with all the other genes of that plant-many of which will dilute the desirable qualities of the original cultivar, or even give undesirable qualities. It can take many years, and five or six back-crosses to the original cultivar to sift out the unwanted genes and recover something resembling the original genotype of the parent.

Much emphasis has been placed on the development of genetically engineered "designer" plants where new genes are added to the plant, or existing genes modified. But while the addition of new genes is significant, the powerful new techniques developed are the real breakthrough. The knowledge gained helps our understanding of the biological basis of conventional plant breeding, and will open the way for a whole new era in plant breeding.

Many genetic techniques are already helping plant breeders to speed up the breeding process. A good example of the interaction between plant breeders and biotechnologists is in the area of developing new pest and disease resistant apples.

Developing New Pest and Disease Resistant Apples

At present most pests and diseases are controlled by applications of chemicals. However, international markets are placing increasingly lower residue limits on imported fruit and there is greater consumer awareness about potential environmental and health issues associated with chemical sprays. In response to these concerns, HortResearch has a diverse research programme aimed at investigating the complete range of options available to replace or supplement chemical sprays to ensure environmentally sustainable supplies of safe, nutritious, affordable food. The most effective long term and environmentally friendly way of reducing spray use is to develop fruit cultivars that are resistant to pests and diseases. Breeding new cultivars by conventional means can be very hit and miss, and take many years. To speed up the process of developing disease resistant cultivars, HortResearch plant breeders and molecular biologists have combined their skills. Genetic manipulation should give the same result more quickly than traditional breeding, but also give greater control. It should mean that the time needed to transfer the natural resistance to apple scab found, for example, in some crab apples to an export apple cultivar like "Gala" or "Braeburn", can be cut in half.

Developing resistant apple cultivars starts with the conventional plant breeder. Freshly germinated apple seedlings, from crosses between susceptible and resistant parents, are exposed to pests and diseases. Information on the resistance or susceptibility of these seedlings to a particular pest or disease, together with leaf samples for DNA analysis, are sent to another group of researchers in HortResearch who are developing an apple gene map. The data provided by the plant breeders helps to locate "DNA markers" which are near each resistance gene on the apple chromosome. (The gene map will also include markers for other important traits.)

These DNA markers will be used in future to provide apple breeders with information on the resistance potential of young seedlings before this potential has been expressed. Such early progeny screening and selection of elite seedlings before they are planted out in research orchards should enable a significant reduction in the requirement for land and plant maintenance, as well as providing breeders with specific information needed to design subsequent crosses. This information on DNA markers will also be used by other molecular biologists to isolate the resistance genes from special "libraries" containing apple DNA. These genes are the key to a range of new disease control methods including transferring genes into other apple cultivars to speed up the breeding process. The whole process is also being helped by HortResearch's development of an extensive collection of old-fashioned apple cultivars which are no longer used commercially, and apple lines collected from around the world both in the wild and from other breeding programmes. This world-first collection will increase the range of genes available both for use in conventional breeding and for genetic engineering.

OTHER USES OF BIOTECHNOLOGY

DNA MARKERS IN KIWIFRUIT

DNA markers are also being developed to assist our kiwifruit breeders. Kiwifruit has separate male and female plants, and at present plants can be distinguished only once they have flowered, usually after several years' growth.

This means that conventional plant breeding programmes are very inefficient as half the seedlings grown are non-productive males. Identification of sex-linked markers will provide a valuable selection tool for breeders in their search for quality female selections. In addition, mapping of markers linked to other fruit-related characters will enable the breeding programmes to be planned and focused more efficiently in all areas, from the initial choice of parents, to the early culling of undesirable phenotypes and the early identification of selections.

Cultivar Identification: "DNA Fingerprinting"

Another use of biotechnology that will benefit all fruit breeders is the development of methods for the identification of clones and cultivars. One method, "DNA fingerprinting", identifies individuals by the unique profile produced when their DNA is separated into a series of fragments and resolved into size classes. The end result is a profile rather similar to the bar code that identifies items in a supermarket. DNA fingerprinting is used widely in forensic science to identify individuals involved a particular crime. Similarly, these DNA profiles can also enable unique cultivars or clones to be identified unequivocally, and the technique has potential in the specific identification of plant cultivars and their subsequent protection.

Disease Detection

Early detection in the field plays a major part in being able to control diseases. Using DNA technology, sensitive detection methods have been developed which can detect the bacteria causing fireblight even when there are no symptoms evident in an orchard. This ability has meant that we are now able to export our apples to Japan.

Alternative Pest and Disease Control Methods

As well as the development of molecular markers for natural resistance genes, HortResearch scientists are trying to find other ways to give plants resistance to pests and diseases.

Fungal Diseases

Root-rotting and fruit-rotting fungi are being studied, using a combination of conventional and molecular techniques, to find ways to disrupt normal infection processes. Investigations are being made of the natural resistance mechanisms operating in plants. These mechanisms control the development of rots, such as production of enzymes which are active against fungal cell walls, or products that inhibit critical fungal enzymes. The genes for these enzymes may be used in developing disease-resistant plants.

Virus Diseases

Biotechnology is also helping in developing control strategies for virus and viroid diseases on many horticultural crops. These agents can adversely affect fruit size, appearance, sugar content, yield and flower break. For example, the tamarillo mosaic virus causes blotchy fruit and leaves, and is the major obstacle to exporting tamarillos. DNA technology is being used to overcome these problems: by splicing the viral coat protein gene into the tamarillo plant, the plant is made resistant to the virus.

Insect Pests

Insect pests are a major cause of crop damage and can also pose quarantine problems on exported fruit. Alternatives to the currently used chemical sprays are urgently required. The bacterium *Bacillus thuringiensis* (Bt) produces a crystal protein which is toxic to selected insects but is completely safe to humans, other species and the environment. Bt is currently used as a bioinsecticide as an alternative to chemical pesticides. Scientists are working towards putting the crystal protein gene into plants to give them "inbuilt" insect resistance.

Fruit Quality

Fruit quality refers to all the factors such as colour, flavour, texture, size and shape, which are the main determinants of fruit acceptability, as well as storage and shelf life. Our ability to control these factors would give us a competitive advantage in export markets. HortResearch scientists are studying many aspects of fruit quality, including colour development and the ripening process.

A method developed in the US on tomatoes, in which scientists insert a mirror-image of the gene involved in fruit softening into the fruit, could be adapted for fruit in this country, particularly where poor storage and shelf life cause problems for exporting the crop. The U.S. studies have found that these genetically engineered fruit were less prone to rotting.

Safety

There have been concerns raised about genetically modified plants escaping into the environment and competing with wild plants. But as has been explained earlier, biotechnology usually only involves the insertion of one or two genes (*i.e.* pieces of DNA) into an existing cultivar. There is less risk involved in this than in conventional plant breeding where a number of undesirable characteristics may show up in crosses. Also, every gene is equipped with a region of the DNA code, called a promoter, which switches the gene on and off, and specifies when, in what cells and in what amount the gene's protein or enzyme is to be made. These "DNA switches" can be tailor-made for specific requirements and can even be externally controlled.

Furthermore, before researchers can even start to experiment in genetic engineering, their plans must be approved by a regulatory body set up in New Zealand to determine which projects are acceptable. All experiments involving transgenic plants must take place within strictly confined glasshouses. Before releasing transgenic plants, it is necessary to carry out a risk assessment to determine whether the transgenic cultivar will behave differently from a conventionally bred cultivar. Assessment procedures are coordinated and regulated internationally by various organisations.

Transgenic plant systems have now been developed worldwide for more than 30 crops as diverse as maize, oats, rice and cotton through to grapes, kiwifruit, papaya and apples. The overwhelming conclusions from nearly five hundred field test experiments on genetically engineered plants in the US and Europe are that newly introduced genes-including those for quality improvement and for control of insects, weeds and plant diseases-are stable, inherited and are expressed like any other plant gene.

Medieval philosopher de Toqueville said "people will believe a simple lie in preference to a complicated truth." For most people, biotechnology is about as complicated as science gets. But biotechnology is something people should be informed about because it is extremely potent in what it can do.

Technical advances are already providing significant commercial benefits to growers, processors and consumers. We believe the growth in the world population and in the demand for food, and the clear consumer preference for environmentally sustainable agriculture will extend biotechnology's role in food production.

The general opinion of the experts is "that biotechnology is simply a better way of doing what breeders have done for millennia in trying to improve agriculture". While biotechnology is unlikely to completely replace conventional plant breeding, it certainly has a place in enabling breeders to achieve their objectives more quickly, safely and accurately. The development of new cultivars with pest and disease resistance and/or improved fruit quality will enable the New Zealand fruit industry to maintain and enhance its competitive advantage in export markets.

GENETICALLY MODIFIED PLANTS

Genetically modified plants are created by the process of genetic engineering, which allows scientists to move genetic material between organisms with the aim of changing their characteristics. All organisms are composed of cells that contain the DNA molecule. Molecules of DNA form units of genetic information, known as genes. Each organism has a genetic blueprint made up of DNA that determines the regulatory functions of its cells and thus the characteristics that make it unique. Prior to genetic engineering, the exchange of DNA material was possible only between individual organisms of the same species. With the advent of genetic engineering in 1972, scientists have been able to identify specific genes associated with desirable traits in one organism and transfer those genes across species boundaries into another organism.

A gene from bacteria, virus, or animal may be transferred into plants to produce genetically modified plants having changed characteristics. Thus, this method allows mixing of the genetic material among species that cannot otherwise breed naturally. The success of a genetically improved plant depends on the ability to grow single modified cells into whole plants. Some plants like potato and tomato grow easily from single cell or plant tissue. Others such as corn, soy bean, and wheat are more difficult to grow.

After years of research, plant specialists have been able to apply their knowledge of genetics to improve various crops such as corn, potato, and cotton. They have to be careful to ensure that the basic characteristics of these new plants are the same as the traditional ones, except for the addition of the improved traits. The world of biotechnology has always moved fast, and now it is moving even faster. More traits are emerging; more land than ever before is being planted with genetically modified varieties of an ever-expanding number of crops. Research efforts are being made to genetically modify most plants with a high economic value such as cereals, fruits, vegetables, and floriculture and horticulture species.

BIOLOGICAL CELL

It only takes one biological cell to create an organism. In fact, there are countless species of single celled organisms, and indeed multi-cellular organisms like ourselves.

A single cell is able to keep itself functional by owning a series of '*miniature machines*' known as *organelles*. The following list looks at some of these organelles and other characteristics typical of a fully functioning cell.

- *Mitochondrion:* An important cell organelle involved in respiration
- *Cytoplasm:* A fluid surrounding the contents of a cell and forms a vacuole
- *Golgi Apparatus:* The processing area for the creation of a glycoprotein
- *Endoplasmic Reticulum:* An important organelle heavily involved in protein synthesis.
- *Vesicles: Packages* of substances that are to be used in the cell or secreted by it.
- *Nucleus*: The "brain" of a cell containing genetic information that determines every natural process within an organism.
- *Cell Membrane:* Also known as a plasma membrane, this outer layer of a cell assists in the movement of molecules in and out the cell plays both a structural and protective role
- *Lysosomes:* Membranous sacs that contain digestive enzymes

CELL WALL

A structure that characteristically is found in plants and prokaryotes and not animals that plays a structural and protective role.

CELL SPECIALISATION

Cells can become specialised to perform a particular function within an organism, usually as part of a larger tissue consisting of many of the same cells working in tandem:

- *Nerve cells* to operate as part of the nervous system to send messages back and forth via the brain at the centre of the nerve system.
- Skin cells for waterproof protection and protection against pathogens in the open air environment.
- *Xylem* tubes to transport water around plants and to provide structural support for the plant as a whole.

Cells combine their efforts in these tissue types to perform a common cause. The task of the specialised cell will determine in what way it is going to be specialised, because different cells are suited to different purposes, as illustrated in the above list:

- Muscle cells are long and smooth in structure and their elastic nature allows these cells to perform flexible movements, just as they do in our own body's.
- Some *white blood cells* contain powerful digestive enzymes to eliminate pathogens by breaking them down to the molecular level.

- Cells at the back of the eye are sensitive to light stimuli, and thus can *interpret* differences in light intensity which can in turn be interpreted by our nervous system and brain.

Many of these cells contain organelles, though after some cells are specialised, they do not possess particular characteristics as they do not require them to be there. *i.e.* efficiency is the key, no resources are wasted and the resources available are put to their idyllic optimum.

THE CELL MEMBRANE

The *cell membrane,* otherwise known as the plasma membrane is a semi-permeable structure consisting mainly of *phospholipid* (fat) molecules and proteins.

They are structured in a *fluid mosaic model,* where a double layer of phospholipid molecules provide a barrier accompanied by proteins. It is present round the circumference of a cell to acts as a barrier, keeping foreign entities out the cell and its contents (like cytoplasm) firmly inside the cell. The plasma membrane allows only selected materials to pass in and out of a cell, and is thus known as a selectively permeable membrane.

CELL TRANSPORT

There are three methods in which ions are transported through the cell membrane into the cell,

- *Active Transport*: Active transport is the transport of molecules with the active assistance of a carrier that can transport the material against a natural *concentration gradient*.
- *Passive Transport (Diffusion):* The movement of molecules from areas of high concentration (*i.e.* outside a cell) to areas of low concentration (*i.e.* within a cell) via a carrier. This process does not require energy.
- *Simple Diffusion:* The movement of molecules from areas of high concentration to areas of low concentration in a free state. *Osmosis* of water involves this type of diffusion through a selectively permeable membrane (*i.e.* plasma membrane)

THE BREAKDOWN OF MATERIALS IN A CELL

In cells, sometimes it is required to breakdown more complex molecules into more simple molecules, which can then be 're-built' into what is needed by the body with these new raw materials. '*Pinocytosis*' where to contents of a structure (such as bacteria) are *drank,* essentially by breaking down molecules into a drinkable form. '*Phagocytosis*' where contents are 'eaten'.

ABSORPTION AND SECRETION

Absorption is the uptake of materials from a cells' external environment. Secretion is the ejection of material.

APPLICATIONS OF PLANT GENETIC ENGINEERING

"PHARMING" AND "PLANTIBODIES"

An increasingly viable option is the production of highly valuable enzymes by plants and animals. In addition to production of human proteins in these organisms, other valuable proteins that are currently produced by microorganisms could very well be produced by higher organisms instead. Animal and plant "bioreactors" in some respects may be superior to recombinant bacterial systems, because eukaryotes glycosylate proteins. Whereas the glycosylation pattern may be species-specific, appropriate glycosylation is often required for protein function. Production through these organismal systems may also be cheaper than cell fermentation techniques.

Two examples of production of human proteins in plants include the production of human serum albumin in transgenic tobacco and potato, and production of human insulin by tobacco. In both cases, the produced protein appears to be fully effective in humans. Unfortunately, however, one cannot raise his/her insulin level by eating transgenic tobacco leaves, as the protein in most cases will be broken down to amino acids before it reaches the blood stream. Therefore, in these cases one cannot escape the practice of protein isolation and purification before transgenic leaves are converted into drugs. Also antibodies are being produced in plants. Initially, the antibody's light and heavy chains were produced in different plants. But a subsequent cross of these two varieties resulted in progeny carrying assembled and functional antibodies.

REVERSIBLE MALE STERILITY IN PLANTS

As has been indicated earlier, heterozygous individuals often are healthier and stronger than homozygous ones. The only way to guarantee heterozygoiscity in plants is to make sure self-pollination cannot occur. For most crop plants it was very tedious or practically impossible to exclude selfing.

To exclude self-pollination, it would be good to introduce male sterility in plants: progeny from such plants are then expected to be 100 per cent heterozygous (assuming they were pollinated with pollen from an unrelated variety). To introduce male sterility, a promoter was identified that was turned on exclusively in tapetum cells (a tissue around the pollen sac that is essential for pollen production).

This promoter then was linked up to a gene coding for a bacterial ribonuclease (named barnase). This ribonuclease selectively chops up ribonucleic acids. The promoter/ribonuclease construct was then introduced into plants (canola, tobacco, you name it). Because the promoter allows expression only in tapetum cells, the gene construct disrupts only development of the tapetal tissue and its end product, pollen. Plants transformed with this construct were male-sterile but otherwise normal. Although male-sterile plants are valuable for hybrid seed production, they have limited value when it comes to crop production. Fertility must be restored to crops such as wheat, rice, and tomato, in which the seed or fruit is the harvested product. Fortunately, the ribonuclease is inhibited very much by a simple protein, named barstar. One can thus cross the male-sterile plant with a male-fertile variety in which the gene for barstar has been introduced, and the result is progeny with viable pollen and restored fertility.

A closely related approach has been criticized as "terminator technology" as it is seen by its critics as a way for companies to protect and enforce their patents.

ANTISENSE RNA

Antisense RNA refers to nucleotide strands that are produced in a cell and that are complementary to a particular mRNA. Antisense RNA can be produced, by inverting the coding region of a gene with respect to its promoter. The antisense RNA can hybridize with its corresponding mRNA, making it double-stranded. The double-stranded mRNA no longer can be recognized by the protein-synthesizing machinery (the ribosomes), and thus expression of this mRNA is suppressed. Also, in many systems double-stranded mRNA is very unstable and is broken down quickly. Thus, one can inactivate specific genes while not interfering with others.

Antisense approaches already are used to protect plants from damage by plant viruses. Reversal of a gene from bean yellow mosaic virus (BYMV), and putting it into tobacco under a reasonably strong promoter, has led to a tobacco variety that is quite resistant to BYMV. A similar approach is used to transfer viral resistance to other plants. This finding is of significance, in that currently no effective, environmentally friendly methods exist to control many plant viruses. Very related to this approach is the RNAi (RNA interference) approach. This is very useful for both agriculture and medicine, and the first examples of practical applications of this RNAi technology are appearing. As with any new technology, the initial pilot projects are sort of pedantic. However, more exciting application possibilities abound. Obviously, RNAi technology provides an excellent approach for reverse genetics in eukaryotes. With this method one can turn off genes.

AGRICULTURAL APPLICATIONS IN DEVELOPING COUNTRIES

Perhaps indicative of the large potential and relative ease of genetic engineering, developing countries (particularly China) are progressing rapidly in development and application of genetically engineered crops. Some have gone into commercial production well ahead of similar crops in the US. In China, tomatoes that have been engineered for improved virus resistance have been on the market since late 1992.

There are two main reasons for the more rapid commercialization of bioengineered crops in the developing world:

- Less tight governmental approval mechanisms,
- Hungrier populations.

While in developed countries the main value of biotechnological applications may be to reduce production costs, in the developing world a main factor is the production of more food. Indeed, genetic engineering applications seem to be pretty successful to cut down on pathogen-induced losses. Genetic modification of papaya plants (expression of the ringspot virus coat protein in the plant) protects very well against the very destructive ringspot virus.

PLANT GENETICS

Knowledge of population genetics, quantitative genetics, probability theory and statistics is indispensable for understanding equilibria and shifts with regard to the genotypic composition of a population, its mean value and its variation. The subject of population genetics is the study of equilibria and shifts of allele and genotype frequencies in populations.

These equilibria and shifts are determined by five forces: Mode of reproduction of the considered crop: The mode of reproduction is of utmost importance with regard to the breeding of any particular crop and the maintenance of already available varieties. This applies both to the natural mode of reproduction of the crop and to enforced modes of reproduction, like those applied when producing a hybrid variety. In plant breeding theory, crops are therefore classified into the following categories: cross-fertilizing crops, self-fertilizing crops, crops with both cross- and self-fertilization and asexually reproducing crops. It is explained that even within a specific population, traits may differ with regard to their mode of reproduction.

- Selection
- Mutation
- Immigration of plants or pollen, *i.e.* immigration of alleles
- Random variation of allele frequencies

A population is a group of (potentially) interbreeding plants occurring in a certain area, or a group of plants originating from one or more common ancestors. The former situation refers to cross-fertilizing crops (in which case the term Mendelian population is sometimes used), while the latter group concerns, in particular, self-fertilizing crops. In the absence of immigration the population is said to be a closed population.

Examples of closed populations are:

- A group of plants belonging to a cross-fertilizing crop, grown in an isolated field, *e.g.* maize or rye (both pollinated by wind), or turnips or Brussels sprouts (both pollinated by insects)

- A collection of lines of a self-fertilizing crop, which have a common origin, *e.g.* a single-cross, a three-way cross, a backcross

The subject of quantitative genetics concerns the study of the effects of alleles and genotypes and of their interaction with environmental conditions. Population genetics is usually concerned with the probability distribution of genotypes within a population (genotypic composition), while quantitative genetics considers phenotypic values (and statistical parameters dealing with them, especially mean and variance) for the trait under investigation.

In fact population genetics and quantitative genetics are applications of probability theory in genetics. An important subject is, consequently, the derivation of probability distributions of genotypes and the derivation of expected genotypic values and of variances of genotypic values. Generally, statistical analyses comprise estimation of parameters and hypothesis testing. In quantitative genetics statistics is applied in a number of ways. It begins when considering the experimental design to be used for comparing entries in the breeding programme.

Considered across the entries constituting a population (plants, clones, lines, families) the expression of an observed trait is a random variable. If the expression is represented by a numerical value the variable is generally termed phenotypic value, represented by the symbol p.

Two genetic causes for variation in the expression of a trait are distinguished. Variation controlled by so-called major genes, *i.e.* alleles that exert a readily traceable effect on the expression of the trait, is called qualitative variation. Variation controlled by so-called polygenes, *i.e.* alleles whose individual effects on a trait are small in comparison with the total variation, is called quantitative variation. In Note: it is elaborated that this classification does not perfectly coincide with the distinction between qualitative traits and quantitative traits.

The former paragraph suggests that the term *gene* and *allele* are synonyms. According to Rieger, Michaelis and Green a gene is a continuous region of DNA, corresponding to one (or more) transcription units and consisting of a particular sequence of nucleotides. Alternative forms of a particular gene are referred to as alleles. In this respect the two terms 'gene' and 'allele' are sometimes interchanged. Thus the term 'gene frequency' is often used instead of the term 'allele frequency'.

The term locus refers to the site, alongside a chromosome, of the gene/allele. Since the term 'gene' is often used as a synonym of the term 'locus', we have tried to avoid confusion by preferential use of the terms 'locus' and 'allele' (as a synonym of the word gene) where possible. In the case of qualitative variation, the phenotypic value p of an entry (plant, line, family) belonging to a genetically heterogeneous population is a discrete random variable. The phenotype is then exclusively (or to a largely traceable degree) a function f of the genotype, which is also a random variable G.

Thus,

$$p = f(G)$$

It is often desired to deduce the genotype from the phenotype. This is possible with greater or lesser correctness, depending for example on the degree of dominance and sometimes also on the effect of the growing conditions on the phenotype.

A knowledge of population genetics suffices for an insight into the dynamics of the genotypic composition of a population with regard to a trait with qualitative variation: application of quantitative genetics is then superfluous.

Note: All traits can show both qualitative and quantitative variation. Culm length in cereals, for instance, is controlled by dwarfing genes with major effects, as well as by polygenes.

The commonly used distinction between qualitative traits and quantitative traits is thus, strictly speaking, incorrect. When exclusively considering qualitative variation, *e.g.* with regard to the traits in pea (*Pisum sativum*) studied by Mendel, this book describes the involved trait as a trait showing qualitative variation.

On the other hand, with regard to traits where quantitative variation dominates – and which are consequently mainly discussed in terms of this variation – one should realise that they can also show qualitative variation.

In this sense the following economically important traits are often considered to be 'quantitative characters':

- Biomass
- Yield with regard to a desired plant product
- Content of a desired chemical compound (oil, starch, sugar, protein, lysine) or an undesired compound
- Resistance, including components of partial resistance, against biotic or abiotic stress factors
- Plant height

In the case of quantitative variation p results from the interaction of a complex genotype, *i.e.* several to many loci are involved, and the specific growing conditions are important. In this book, by complex genotype we mean the sum of the genetic constitutions of all loci affecting the expression of the considered trait.

These loci may comprise loci with minor genes (or polygenes), as well as loci with major genes, as well as loci with both. With regard to a trait showing quantitative variation, it is impossible to classify individual plants, belonging to a genetically heterogeneous population, according to their genotypes.

This is due to the number of loci involved and the complicating effect on p of (some) variation in the quality of the growing conditions. It is, thus, impossible to determine the number of plants representing a specified complex genotype. (With regard to the expression of qualitative variation this may be possible!). Knowledge of both population genetics and quantitative genetics is therefore required for an insight into the inheritance of a trait with quantitative variation.

The phenotypic value for a quantitative trait is a continuous random variable and so one may write:

$$p = f(G,e)$$

Thus the phenotypic value is a function f of both the complex genotype (represented by G) and the quality of the growing conditions (say environment, represented by e). Even in the case of a genetically homogeneous group of plants (a clone, a pure line, a single-cross hybrid) p is a continuous random variable.

The genotype is a constant and one should then write:

$$p = f(G,e)$$

Regularly in this book, simplifying assumptions will be made when developing quantitative genetic theory.

Especially the following assumptions will often be made:

- Absence of linkage of the loci controlling the studied trait(s)
- Absence of epistatic effects of the loci involved in complex genotypes.

These assumptions will now be considered.

Absence of Linkage

The assumption of absence of linkage for the loci controlling the trait of interest, *i.e.* the assumption of independent segregation, may be questionable in specific cases, but as a generalisation it can be justified by the following reasoning. Suppose that each of the n chromosomes in the genome contains M loci affecting the considered trait. This implies presence of n groups of,

$$\binom{M}{2} -$$

pairs of loci consisting of loci which are more strongly or more weakly linked. The proportion of pairs consisting of linked loci among all pairs of loci amounts then to,

$$\frac{n\binom{M}{2}}{\binom{nM}{2}} = \frac{n.M!}{2!(M-2)!} \times \frac{2!(nM-2)!}{(nM)!} = \frac{M-1}{nM-1} = \frac{1-\frac{1}{M}}{n-\frac{1}{M}}$$

For $M = 1$ this proportion is 0; for $M = 2$ it amounts to 0.077 for rye (*Secale cereale,* with $n = 7$) and to 0.024 for wheat (*Triticum aestivum,* with $n = 21$); for $M = 3$ it amounts to 0.100 for rye and to 0.032 for wheat. For $M \rightarrow \infty$ the proportion is 1 n; *i.e.* 0.142 for rye and 0.048 for wheat.

One may suppose that loci located on the same chromosome, but on different sides of the centromere, behave as unlinked loci. If each of the n chromosomes contains,

$$m\left(=\frac{1}{2}M\right)$$

relevant loci on each of the two arms then there are $2n$ groups of,

$$\binom{m}{2}-$$

pairs consisting of linked loci. Thus considered, the proportion of pairs consisting of linked loci amounts to,

$$\frac{2n\binom{m}{2}}{\binom{2nm}{2}}=\frac{2n.m!}{2!(m-2)!}\times\frac{2!(2nm-2)!}{(2nm)!}=\frac{1-\frac{1}{m}}{2n-\frac{1}{m}}$$

For $m = 1$ this proportion is 0; for $m = 2$ it amounts to 0.037 for rye and to 0.012 for wheat; for $m = 3$ it amounts to 0.049 for rye and to 0.016 for wheat. For $m \rightarrow \infty$ the proportion is,

$$\frac{1}{2n}$$ .e. 0.071 for rye and 0.024 for wheat.

For the case of an even distribution across all chromosomes of the polygenic loci affecting the considered trait it is concluded that the proportion of pairs of linked loci tends to be low. (In an autotetraploid crop the chromosome number amounts to $2n = 4x$. The reader might like to consider what this implies for the above expressions.)

ABSENCE OF EPISTASIS

Absence of epistasis is another assumption that will be made regularly in this book. It implies additivity of the effects of the single-locus genotypes for the loci affecting the level of expression for the considered trait.

The genotypic value of some complex genotype consists then of the sum of the genotypic value of the complex genotype with regard to all non-segregating loci, here represented by *m*, as well as the sum of the contributions due to the genotypes for each of the *K* segregating polygenic loci *B*1-*b*1..., *BK*-*bK*. Thus,

$$GB_1-b_1....,B_k-b_k=m+G'B_1-b_1+...+G'B_k-b_k$$

where G_ is defined as the contribution to the genotypic value, relative to the population mean genotypic value, due to the genotype for the considered locus.

The assumption implies the absence of inter-locus interaction, *i.e.* the absence of epistasis (in other words: absence of non-allelic interaction).

It says that the effect of some genotype for some locus *Bi-bi* in comparison to another genotype for this same locus does not depend at all on the complex genotype determined by all other relevant loci. In this book, in order to clarify or substantiate the main text, theoretical examples and results of actual experiments are presented. Notes provide short additional information and appendices longer, more complex supplementary information or mathematical derivations.

PHOTOSYNTHESIS-PHOTOLYSIS AND CARBON FIXATION

Photosynthesis is the means that primary producers (mostly plants) can obtain energy via light energy. The energy gained FROM light can be used in various processes mentioned below for the creation of energy that the plant will need to survive and grow.

Photosynthesis is a reduction process, where hydrogen is reduced by a coenzyme.

This is in contrast to respiration where glucose is oxidised. The process is split INTO two DISTINCT areas, *photolysis* (the photochemical stage) and the *Calvin Cycle*. A summary of the reaction, where light energy is used to initiate the reaction in its presence;

$$CO_2 + H_2O \rightarrow \text{glucose} + \text{oxygen}$$

PHOTOLYSIS

This part of photosynthesis occurs in the *granum* of *a chloroplast* where light is absorbed by *chlorophyll*; a type of photosynthetic pigment that converts the light to chemical energy. This reacts with water (H_2O) and splits the oxygen and hydrogen molecules apart. From this dissection of water, the oxygen is released as a by-product while the reduced hydrogen acceptor makes its way to the second stage of photosynthesis, the Calvin cycle. Overall, since the water is oxidised (hydrogen is removed) and energy is gained in photolysis which is required in the Calvin cycle.

THE CALVIN CYCLE

Also known as the carbon fixation stage, this part of the photosynthetic process occurs in the *stroma* of chloroplasts. The carbon made available FROM breathing in carbon dioxide enters this cycle: Just like the *Kreb's Cycle* in respiration, a substrate is manipulated INTO various carbon compounds to produce energy. In the case of photosynthesis, the following steps occur, which create glucose for respiration FROM the carbon dioxide introduced INTO the cycle;

- Carbon FROM CO_2 enters the cycle combining with Ribulose Biphosphate (RuBP)
- A compound formed is unstable and breaks down FROM its 6 carbon nature to a 3 carbon compound called glycerate phosphate (GP)
- Energy is used to break down GP INTO triose phosphate, while a hydrogen acceptor reduces the compound therefore requiring energy

- Triose Phosphate is the end product of this, a 3 carbon compound which can double up to form glucose, which can be used in respiration.
- The cycle is completed when the leftover GP molecules are met with a carbon acceptor and then turned INTO RuBP, which is to be joined with the carbon dioxide molecules to re-begin the process.

The energy that is used up in the Calvin cycle is the energy that is made available during photolysis. The glucose that is made via GP can be used in respiration or a building block in forming *starch* and *cellulose,* materials that are commonly in demand in plants.

LIMITING FACTORS IN PHOTOSYNTHESIS

Some factors affect the rate of photosynthesis in plants, as follows:

- *Temperature* plays a role in affecting the rate of photosynthesis. Enzymes involved in the photosynthetic process are directly affected by the temperature of the organism and its environment.
- Light Intensity is also a *limiting factor,* if there is no sunlight, then the photolysis of water cannot occur without the light energy required.
- Carbon Dioxide concentration also plays a factor, due to the supplies of carbon dioxide required in the Calvin cycle stage.

Overall, this is how a plant produces energy which supplies a rich source of glucose for respiration and the building blocks for more complex materials. While animals get their energy FROM food, plants get their energy FROM the sun.

CELL RESPIRATION

As mentioned in the previous page on ATP, the process of *respiration* is split into 3 distinct areas that occur at different parts of the cell. Respiration involves the *oxidation* of foodstuff (*i.e.* glucose) in order to create ATP. Respiration can occur with or without oxygen, *aerobic* and *anaerobic* respiration respectively.

GLYCOLYSIS

Glycolysis occurs in the *cytoplasm* of a cell where a 6 carbon glucose molecule (the broken down food that you ate earlier) is broken down by enzymes into a 3 carbon *pyruvic acid*. The execution of this process requires 2 ATP, and produces a net gain of 2 ATP. The enzymes involved remove hydrogen from the glucose (oxidation) where they take these hydrogen atoms to the cytochrome system, explained soon.

In anaerobic respiration, this is where the process ends, glucose is split into 2 molecules of pyruvic acid. When oxygen is present, pyruvic is broken down into other carbon compounds in the Kreb's Cycle. When it is not present, the pyruvic acid is broken down into lactic acid (or carbon dioxide and ethanol).

THE KREB'S CYCLE

When oxygen is present, respiration can harness more ATP from a single unit of glucose. The *pyruvic acid* from the *glycolysis* stage diffuses into a cell organelle called a mitochondrion (pl. *mitochondria*).

These mitochondria are sausage shaped structures that host a large surface area for the respiration to occur on.

The pyruvic acid is then subject to more enzymes which break it down into a 2 carbon compound.

The diagram illustrates the Kreb's cycle, consisting of three main actions:

- The carbon element is in an infinite cycle where the 2 carbon compound derived from pyruvic acid binds with the 4 carbon compound that is always present in the cycle.
- CO_2 is released, where the oxygen that is present in *aerobic respiration* combines with carbon from the carbon compounds which is released as CO_2. Hence the need for animals to breath out and expel this CO_2.
- Enzymes oxidize the carbon compounds and transport the hydrogen atoms to the cytochrome system.

THE CYTOCHROME SYSTEM

The cytochrome system, also known as the hydrogen carrier system (or the *electron transport system*) are where the reduced hydrogen carriers transport hydrogen atoms from the glycolysis and Kreb's cycle stages. The cytochrome system is found in the many *cristae* of mitochondria, which are tiny stalked particles found on its outer layer.

The system contains many 'hydrogen acceptors' which hydrogen can be added to. By following the path of a hydrogen atom, we can see how the cytochrome system works:

- Some *coenzymes* from earlier stages are transferred to the next coenzymes.
- B is then oxidised, therefore the coenzyme releases the hydrogen and energy is made available.
- The released hydrogen atom binds with 2 oxygen atoms (oxygen is available in aerobic respiration) which produces water, a by-product of respiration.

The diagram illustrates this flow of hydrogen within the cytochrome system and how energy is made available by the flow of these atoms. The green circles illustrate where energy is made available via oxidation. Overall their is a gain of 38 ATP from one molecule of glucose in aerobic respiration.

The food that we eat provides glucose required in respiration. In plants, energy is also acquired via respiration, but the mechanism of delivering glucose to the respiration process is a little different.

BIOLOGICAL ENERGY-ADP AND ATP

ATP stands for Adenosine Tri-Phosphate, and is the energy used by an organism in its daily operations. It consists of an *adenosine* molecule and three inorganic *phosphates*.

After a simple reaction breaking down ATP to *ADP*, the energy released from the breaking of a molecular bond is the energy we use to keep ourselves alive.

ATP TO ADP-ENERGY RELEASE

This is done by a simple process, in which one of the phosphate molecules is broken off, therefore reducing the ATP from 3 phosphates to 2, forming ADP (Adenosine Diphosphate after removing one of the phosphates {Pi}). This is commonly wrote as ADP + Pi.

When the bond connecting the phosphate is broken, *energy* is released. While ATP is constantly being used up by the body in its biological processes, the energy supply can be bolstered by new sources of glucose being made available via eating food which is then broken down by the digestive system to smaller particles that can be utilised by the body.

On top of this, ADP is built back up into ATP so that it can be used again in its more energetic state. Although this conversion requires energy, the process produces a net gain in energy, meaning that more energy is available by re-using ADP+Pi back into ATP.

GLUCOSE AND ATP

Many ATP are needed every second by a cell, so ATP is created inside them due to the demand, and the fact that organisms like ourselves are made up of millions of cells.

Glucose, a sugar that is delivered via the bloodstream, is the product of the food you eat, and this is the molecule that is used to create ATP. Sweet foods provide a rich source of readily available glucose while other foods provide the materials needed to create glucose.

This glucose is broken down in a series of *enzyme* controlled steps that allow the release of energy to be used by the organism. This process is called respiration.

RESPIRATION AND THE CREATION OF ATP

ATP is created via respiration in both animals and plants. The difference with plants is the fact they attain their food from elsewhere. In essence, materials are harnessed to create ATP for biological processes. The energy can be created via cell respiration.

The process of respiration occurs in 3 steps (when oxygen is present:

- Glycolysis
- The Kreb's Cycle
- The Cytochrome System

DNA STRUCTURE AND DNA REPLICATION

The energy required by these cells and how energy is created in order for the cell to survive.The structure, type and functions of a cell are all determined by *chromosomes* that are found in the *nucleus* of a cell. These chromosomes are composed of DNA, the acronym for deoxyribonucleic acid.

This DNA determines all the characteristics of an organism, and contains all the genetic material that makes us who we are.

This information is passed on from generation to generation in a species so that the information within them can be passed on for the offspring to harness in their lifetime.

STRUCTURE OF DNA AND NUCLEOTIDES

DNA is arranged into a *double helix* structure where spirals of DNA are intertwined with one another continuously bending in on itself but never getting closer or further away. The following diagram illustrates a *nucleotide*, the building blocks of DNA:

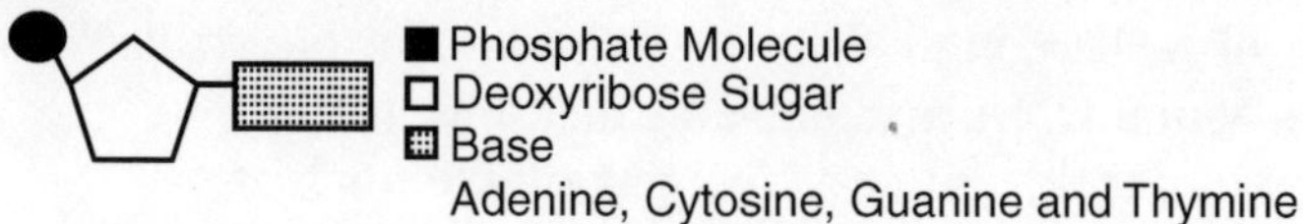

There are four different types of nucleotide possible in a DNA sequence, adenine, cytosine, guanine and thymine (can be replaced with A, C, G and T). There are billions of these nucleotides in our genome, and with all the possible permutations; this is what makes us unique. Nucleotides are situated in adjacent pairs in the double helix nature mentioned. The following rules apply in regards to what nucleotides pair with one another.

- There are four possible types of nucleotide, adenine, cytosine, guanine and thymine.
- Thymine and adenine can only make up a base pair
- Guanine and cytosine can only make up a base pair
- Therefore, thymine and cytosine would NOT make up a base pair, as is the case with adenine and guanine.

This is illustrated in the below diagram, using correct pairings of nucleotides

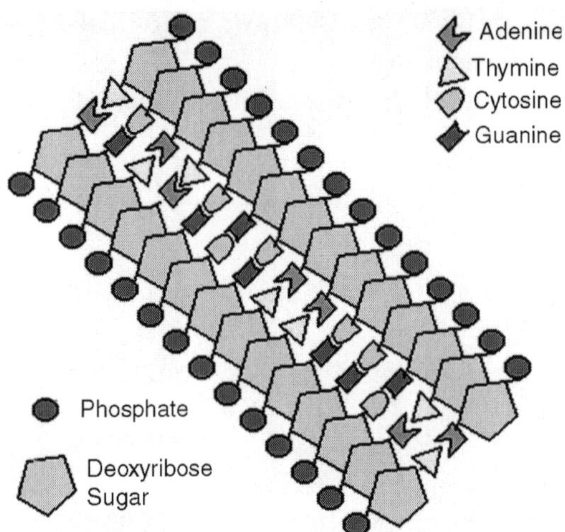

The diagram is two dimensional, remember that DNA is structured in a double helix fashion. This continuous sequence, and the sequence they are in determine an organisms' structural, physical and anatomical features.

DNA REPLICATION

Cells do not live forever, and in light of this, they must pass their genetic information on to new cells, and be able to replicate the DNA to be passed on to offspring.

It is also required that fragments of DNA (genes) have to be copied to code for particular bodily function. It is essential that the replication of it is EXACT.

In order for replication to occur, the following must be available:

- The actual DNA to act as an exact template
- A pool of relevant and freely available nucleotides
- A supply of the relevant enzymes to stimulate reaction
- ATP to provide energy for these reactions

When replicating, the double helix structure uncoils so that each strand of DNA can be exposed.

When they uncoil, the nucleotides are exposed so that the freely available nucleotides can pair up with them.

When all nucleotides are paired up with their new partners, they re-coil into the double helix. As there are two strands of DNA involved in replication, the first double helix produces 2 copies of itself via each strand.

It is said that the replicated DNA is semi-conservative, because it possesses 50percent of the original genetic material from its parent. These 2 new copies have the exact DNA that was in the previous one. This template technique allows genetic information to be passed from cell to cell and from parents to offspring.

PROTEIN SYNTHESIS

If you have jumped straight to this page, you may wish to look at the previous page about DNA, which gives background information on protein synthesis. As mentioned, a string of *nucleotides* represent the genetic information that makes us unique and the blueprint of who and what we are, and how we operate. Part of this genetic information is devoted to the synthesis of *proteins,* which are essential to our body and used in a variety of ways. Proteins are created from templates of information in our DNA:

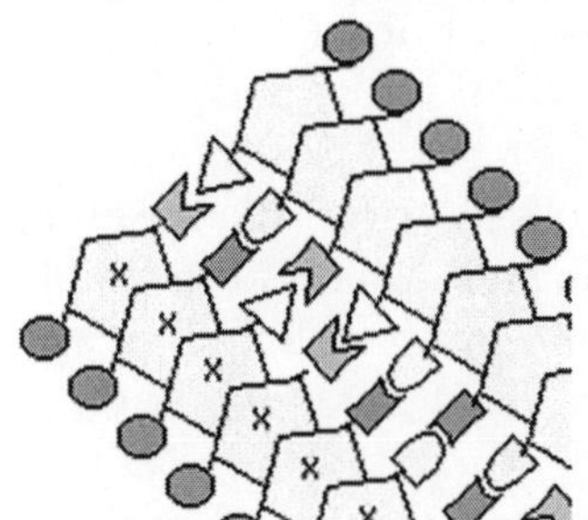

The X marked nucleotides are an example of a DNA sequence that would be used to code for a particular protein, with the sequence of these nucleotides determining which protein it is. The sequence of these nucleotides are used to create amino acids, where chains of *amino acids* form to make a protein.

MRNA

This genetic information is found in the nucleus, though protein synthesis actually occurs in *ribosomes* found in the *cytoplasm* and on rough endoplasmic reticulum. If protein is to be synthesised, then the genetic information in the *nucleus* must be transferred to these ribosomes. This is done by *mRNA* (messenger ribonucleic acid). It is very similar to DNA, but fundamentally differs in two ways:

- A base called *uracil replaces* all thymine bases in mRNA.
- The deoxyribose sugar in DNA in is *replaced* by ribose sugar in mRNA.

At the beginning of protein synthesis, just like DNA replication, the double helix structure of DNA uncoils in order for mRNA to replicate the genetic sequence responsible for the coding of a particular protein. In the beginning, the DNA has uncoiled, allowing the mRNA to move in and transcribe (copy) the genetic information. If the code of DNA looks like this: G-G-C-A-T-T, then the mRNA would look like this C-C-G-U-A-A (remembering that uracil replaces thymine) With the genetic information responsible for creating substances now available on the mRNA strand, the mRNA moves out of the nucleus and away from the DNA towards the ribosomes.

PROTEIN VARIETY

As mentioned in the previous two pages investigating protein synthesis, each consists of a successive chain of amino acids. The sequence of these amino acids determine which type of protein it is. It is synthesised from a DNA strand, each DNA strand involved in protein synthesis is responsible for producing a unique protein.

TYPES OF PROTEIN

Over time and diversity of organisms, a huge amount of proteins exist and perform a unique function in the body.

Primarily, their are three types of protein:

1. *Fibrous Proteins*: These fibre like proteins are used for structural purposes in organisms. This is because fibrous proteins are arranged in long strands and are insoluble in water. Examples of use include providing a barrier in the cell wall of plants and myosin in skeletal muscle
2. *Globular Proteins*: The polypeptide chains (protein chains) in globular proteins are folded together into a knot like shape essential in the fact that are present in the following
   - *Enzymes*: Biological catalysts, enzymes are responsible speeding up reactions in an organism
   - *Hormones*: Hormones are chemical messengers responsible for initialising a response in organisms. Some hormones have a regulatory effect,
   - *Antibodies*: Antibodies are used to defend the body against foreign agents *e.g.* bacteria, fungi and viruses. The next page investigates these.
   - *Structural Protein*: Globular proteins form part of the cell membrane, which has a structural role as well as a role in transporting ions in and out the cell.
3. *Conjugated Proteins*: Conjugated proteins are essentially globular proteins that possess non-living substances, such as the haem found in haemoglobin, which possesses iron (a non-living substance)

Therefore proteins play a vital role in many of an organisms biological processes and their organs. The following page investigates cell defence against foreign agents, where proteins are playing their role in the form of antibodies.

ROLE OF GOLGI APPARATUS AND ENDOPLASMIC RETICULUM IN PROTEIN SYNTHESIS

Continued from the previous page that introduces protein synthesis...

MRNA AND TRNA

mRNA leaves the nucleus and enters the *cytoplasm* where ribosomes can be found, the site of protein synthesis. The mRNA strand is met in the ribosome by complimentary tRNA anticodons, which have opposing bases to that of the mRNA strand (the codons). If the mRNA sequence is A-A-U-C-A-U, (codon) then the tRNA sequence is U-U-A-G-U-A (anticodon) Each *tRNA* molecule consists of 3 bases, deemed an *anticodon* which compliments the opposing bases on the mRNA strand. These in turn have the *amino acid* sequence to successfully code for a particular amino acid. Each amino acid has a certain sequence of *bases* to make it unique. *Therefore, as a summary*:

- The initial DNA contained a certain sequence of nucleotides
- The mRNA has a pre-determined sequence (because it is transcribed from the DNA)
- Again, as a consequence the anticodons possess a pre-determined sequence due to the mRNA
- As each amino acid corresponds to a particular anticodon, a unique amino acid sequence is created forming a protein

These amino acids (*peptides*) can combine to form a *polypeptide* chain (proteins), which are used in a variety of structures such as enzymes and hormones.

RIBOSOMES AND ROUGH ENDOPLASMIC RETICULUM (RER)

Ribosomes are the site of protein synthesis, and can occur freely in the cytoplasm though more commonly on the outer surface of rough endoplasmic reticulum. The *endoplasmic reticulum* presents a large surface area on which these ribosomes can be situated, therefore allowing protein synthesis to occur on a large scale. Rough endoplasmic reticulum is particularly abundant in growing cells which demand a high turnover of materials in its growth. Rough ER is responsible for transporting the newly synthesised proteins to the Golgi apparatus.

THE GOLGI APPARATUS

The *Golgi apparatus* is composed of flattened fluid-filled sacs that controls the flow of molecules in a cell. This is also the case of protein. Carbohydrates are added to the protein to complete its production. This finished product, glycoprotein, is 'pinched off' the Golgi apparatus, and is transported by a vesicle of the cell membrane. When this vesicle reaches the cell membrane, it binds to a receptor on the surface and excretes the protein, where it can then undergo its function.

BIOLOGICAL VIRUSES

The prime directive of all organisms is to reproduce and survive, which is also the case for viruses, which in most cases are considered a nuisance to humans.

VIRUSES

Viruses possess both living and non-living characteristics. The unique characteristic that differentiates viruses from other organisms is the fact that they require other organisms to host themselves in order to survive, hence they are deemed *obligate parasites.*

Viruses can be spread in the following exemplar ways:

- *Airborne*: Viruses that infect their hosts from the open air
- *Blood Borne*: Transmission of the virus between organisms when infected blood enters an organisms circulatory system
- *Contamination*: Caused from the consumption of materials by organisms such as water and food which have viruses within

Therefore viruses have many means of getting transmitted from one organism to another.

CELL ASSIMILATION BY A VIRUS

Viruses are tiny micro-organisms, and due to their size and simplicity, they are unable to replicate independently. Therefore, when a virus is situated in a host, it requires the means to reproduce before it dies out without producing more viruses. This is done by altering the genetic make up of a cell to start coding for materials required to make more viruses. By altering the cell instructions, more viruses can be produced which in turn, can affect more cells and continue their existence as a species. The following is a step by step guide of how an example bacteriophage (a virus that infects bacteria) takes control of its host cell and reproduces itself.

- The virus approaches the bacteria and attaches itself to the cell membrane
- The tail gives the virus the means to thrust its genetic information into the bacteria
- Nucleotides from the host are 'stolen' in order for the virus to create copies of itself
- The viral DNA alters the genetic coding of the host cell to create protein coats for the newly create viral DNA strands
- The viral DNA enters its DNA coat
- The cell is swollen with many copies of the original virus and bursts, allowing the viruses to attach themselves to other nearby cells
- The process begins all over again with many more viruses attacking the hosts' cells

Without a means of defence, the host that is under attack from the virus would soon die.

BIOLOGICAL CELL DEFENCE

Organisms must find a means of defence against antigens such a viruses described on the previous page. If this was not the case, bacteria, fungi and viruses would replicate out of control inside other organisms which would most likely already be extinct.

Therefore organisms employ many types of defence to stop this happening. Means of defence can be categorised into first and second lines of defence, with the first line usually having direct contact with the external environment.

First Lines of Defence

- *Skin* is an excellent line of defence because it provides an almost impenetrable biological barrier protecting the internal environment.
- *Lysozyme* is an enzyme found in tears and saliva that has powerful digestive capabilities, and can break down foreign agents to a harmless status before they enter the body.
- The clotting of blood near open wounds prevents an open space for antigens to easily enter the organism by coagulating the blood.
- *Mucus* and *cilia* found in the nose and throat can catch foreign agents entering these open cavities then sweep them outside via coughing, sneezing and vomiting.
- The *cell wall* of plants consists of fibrous proteins which provide a barrier to potential parasites (antigens).

If these first lines of defence fail, then there are further defences found within the body to ensure that the foreign agent is eliminated.

Second Lines of Defence

Second lines of defence deal with antigens that have bypassed the first lines of defence and still remain a threat to the infected organism. *Interferons* are a family of proteins that are released by a cell that is under attack by an antigen. These interferons attach themselves to *receptors* on the plasma membrane of other cells, effectively instructing it of the previous cells' situation. This tells these neighbouring cells that an antigen is nearby and instructs them to begin coding for antiviral proteins, which upon action, defend the cell by shutting it down. In light of this, any invading antigen will not be able to replicated its DNA (or mRNA) and *protein coat* inside the cell, effectively preventing the spread of it in the organism. These antiviral proteins provide the organism with protection against a wide range of viruses.

This action brought about by interferon is a defensive measure, while *white blood cells* in the second line of defence in animals can provide a means of attacking these antigens. One method of attacking antigens is by a method called *phagocytosis*, where the contents of the antigen are broken down by molecules called phagocytes. These phagocytes contain digestive enzymes in their lysosomes (an organelle in phagocytes) such as lysozyme. White blood cells such as a *neutrophil* or a monocyte are capable of undergoing phagocytosis.

- The bacterium inside the cell gives out chemical messages that are picked up by the phagocyte.
- The bacteria targets the cell as a possible host and moves towards it.
- The cell is prepared for this and the bacterium becomes trapped in a vacuole that forms around it.
- The bacterium is a sitting duck that is harmless at present.
- The lysosomes detect the bacterium and the digestive enzymes inside them begin to break the bacterium down.
- The remnants of the lysosome and bacterium materials are absorbed into the cytoplasm.

PASSIVE AND ACTIVE TYPES OF IMMUNITY

The previous page investigated the role of white blood cells in phagocytosis. *White blood cells* are also responsible for *antibody* formation. Certain antibodies are synthesised in response to the presence of certain *antigens*

Specific Immune Responses

Lymphocytes are a type of white blood cell capable of producing a *specific immune response* to unique antigens. Some of these lymphocytes are capable of entrapping antigens on their surface.

When lymphocytes catch these antigens they can then begin to code for unique antibodies, structures that are capable of catching these antigens. The lymphocytes code for a particular antibody on response to a particular antigen. The antibody that is formed will be capable of catching free antigens therefore neutralising.

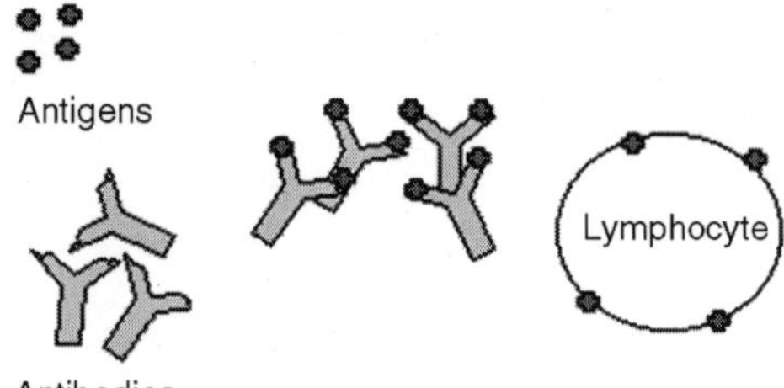

B lymphocytes (B Cells) produce free moving antibodies as above while *T lymphocytes* (T Cells) produce antibodies on their surface.

Types of Immunity

When attacked an organism has several means in which it can prepare to defend itself in event of attack.

- *Active Immunity*: *Vaccines* are used for health purposes to expose our bodies to a particular antigen. These antigens are usually killed or severely weakened to decrease their potency. After destroying these pathogens, the body stores some T cells as memory cells, due to the fact they code for a particular antigen and can be when needed. This memory in T cells can be a means of artificially acquiring immunity while a genuine attack by a pathogen is a naturally acquired type of immunity.

- *Passive Immunity*: This is where immunity to particular antigens as a result of genetic traits passed on from parents rendering the offspring immune to a particular pathogenic threat.

PLANT CELL DEFENCE

HYDROGEN PEROXIDE

Plants release *hydrogen peroxide* in response to the presence of a fungal invasion, which attacks by piercing the cell wall of a plant and breaking it down. This hydrogen peroxide (chemical symbol H_2O_2) is a double edged sword in its defence against the antigen.

One Way: Hydrogen peroxide stops the breakdown of the cell wall Certain pathogens will use pectinase, a digestive enzyme, to break down the cell wall barrier and invade the plant.

The pectinase released by the fungus must be stopped. H_2O_2 *is involved in halting the action of this pectinase*:

- The H_2O_2 is created and moves to the cell wall-the site of the invasion.
- It reacts in contact with and enzyme called *peroxidase*, which promotes the breakdown of *pectinase.*
- The foreign chemical is rendered useless.
- Threat of the cell wall being compromised is removed.

Another Way: Some of the H_2O_2 triggers the creation of phytoalexins

Phytoalexins are similar to the antiviral proteins previously mentioned as a secondary line of defence. Phytoalexins are a family of hormones that inhibit protein synthesis and thus "shut up shop" in the event of a pathogenic attack by halting the protein production process in our cells.

- Chemicals released by the fungus that are being used by it in its attack trigger a chemical response in the plasma membrane that makes the plant aware of the pathogens presence.
- Hydrogen peroxide from the plasma membrane triggers a chemical response to inform the nucleus of the infected cells of the current situation.
- mRNA from the nucleus is transported to ribosomes, as described in the protein synthesis section, where phytoalexins are to be produced (essentially protein synthesis coding for phytoalexins).
- The phytoalexins then take on a role similar to that of antiviral proteins, where the presence of a phytoalexin in a cell inhibits

protein synthesis and therefore preventing growth of the foreign agent by removing all possible avenues of invasion for the pathogen, thus eliminating the threat.

BARRIERS USED BY PLANTS IN DEFENCE

Lignin is a strong type of molecule that provides plants with a defensive structure similar to that of fibrous proteins. It acts as a barrier and can be found in wood and is characteristically found in plants that have recently endured pathogen attack. *Callose* seals off sieve plates in the plant, effectively shutting off the transport of molecules around the organism.

This is done to minimise the chance of the plant transporting infectious material around its own self, and halting the movement of materials that could be used by the pathogen in aid of replicating itself.

Ethylene promotes leaf abscission, and is done to sever the plant of dead or dying plant matter. This is done to prevent the spread of infected material, therefore sacrificing infected sections of plant is more economical than taking the risk of the infection spreading. *Galls* and *tannins* are created by the plant to encapsulate foreign agents found within the plant. A gall is an instance where an infected cell becomes inflamed that contains tannins. These tannins play a protective role by segregating the foreign agent and its chemicals from the rest of the plant

All of the four previous pages have illustrated means of self defence against *pathogens* (fungi, viruses and bacteria).

GENETIC EFFECTS OF INBREEDING

Inbreeding occurs if mating plants are, on the average, *more* related than random pairs of plants. A more than average relatedness of the mating plants is thus a prerequisite. Relatedness implies, of course, that the plants involved share one or more ancestors. The strength of the inbreeding depends on the degree of relatedness of the mating plants. It has already been noted that mating of related plants may occur in random mating, but in that case it occurs as a matter of chance.

Note: Several yardsticks for measuring the degree of relatedness exist, a common one being the probability that an allele of a certain locus in some plant is identical by descent to an arbitrary allele at that same locus in its mate. In regular systems of inbreeding the degree of relatedness of the mating plants is uniform across all pairs of mating plants. In this book no attention is given to the determination of the degree of relatedness.

Regular systems of inbreeding are far more common in plant breeding than irregular systems. No attention will, therefore, be given to irregular systems of inbreeding. The counterpart of inbreeding is outbreeding. With outbreeding mating plants are on the average *less* related than random pairs of plants. Selfincompatibility is a natural cause for outbreeding as related plants tend to have a similar genotype at the incompatibility locus/loci. After intercrossing, such plants will produce no (or few) offspring.

Artificial forms of outbreeding are:

- Bulk crossing of two unrelated populations
- Selection of parents to be crossed in such a way that inbreeding is avoided as much as possible Outbreeding occurs also in the case of immigration.

The population genetic effect of inbreeding is a decrease in the frequency of heterozygous plants. This involves all loci, for all traits. (Random mating, on the other hand, is a mode of reproduction that may occur for certain traits and may simultaneously be absent for other traits). When starting with an F2 population and considering segregating loci, the frequency of heterozygous plants is the same for all loci. This applies to the successive generations of the superpopulation. Each subpopulation consists of few plants: in the case of selfing only a single plant, in the case of full sib mating only pairs of plants. Within these separate subpopulations reproduction is by means of random mating. The random variation of the gene frequencies occurring in small populations causes the subpopulations to vary with regard to the frequencies of heterozygous plants: not only for different loci, but also for the same locus. Individual plants of the F2 (or F3, etc.) populations vary therefore in the number of heterozygous loci. In diploid crops procedures for the production of doubled haploid lines (DH-lines) allow the production of pure lines from heterozygous parents in a single generation.

Doubling of the number of chromosomes of haploid plants, generated by parthenogenesis or by anther culture, yields immediately complete homozygosity. For dioecious crops as well as for self-fertilizing crops with a long juvenile phase, *e.g. Coffea arabica* L., this approach is an attractive alternative to continued inbreeding. Tissue culture techniques for the regeneration of plants from anthers or microspores have been developed, for example in wheat, barley, rice and oilseed rape. Also elimination of paternal chromosomes, occurring when making *Hordeum vulgare* L. × *H. bulbosum* L. or *Triticum aestivum* L. × *Zea mays* L. crosses, permits production of DH-lines. (The paternal chromosomes are lost in a few cell divisions of the hybrid zygote/embryo.)

Note: DH-lines are mostly obtained directly from the gametes produced by the F1-plants. This has a few drawbacks

- Recombination is restricted to the F1 meiosis
- The proportion of DH-lines that are rejected because of poor performance is high. This is undesirable because of the cost of producing DH-lines.

To avoid these drawbacks one may use gametes from plants obtained by backcrossing the F1 or one may use F2- or even F3-plants. (The latter allows selection among F2-plants, followed by selection among F3-lines in the seedling stage). *In vitro* selection among the haploid embryos appeared to be feasible: the size and degree of embryo differentiation predicted which embryos would produce vigourous seedlings. Additionally the growth rate of the embryos was positively correlated with yield performance in the field $r = 0.3$, but this has found little practical application).

Continued self-fertilization is the natural mode of reproduction of selffertilizing crops. There are many economically important self-fertilizing crops.

A number of these are:

- Barley *Hordeum vulgare* L.
- Oats *Avena sativa* L.
- Wheat *Triticum aestivum* L.
- Rice *Oryza sativa* L.
- Sorghum *Sorghum bicolour* (L.) Moench.
- Finger millet *Eleusine coracana* (L.) Gaertn.
- Pea *Pisum sativum* L.
- Cowpea *Vigna unguiculata* (L.) Walp.
- Dry bean *Phaseolus vulgaris* L.
- Soybean *Glycine max* (L.) Merr.
- Peanut *Arachis hypogaea* L.
- Cotton *Gossypium* spp.
- Arabica coffee *Coffea arabica* L.
- Lettuce *Lactuca sativa* L.
- Tomato *Lycopersicon esculentum* Mill.
- Okra *Abelmoschus esculentus* (L.) Moench.
- Sweet pepper *Capsicum annuum* L.

Self-fertilization is not always 100% in most of these autogamous crops, *e.g.* cotton, okra, sorghum. (The amount of outcrossing in sorghum is about 6%.) consider the genotypic composition of populations reproducing by a mixture of self-fertilization and cross-fertilization. Breeders regularly apply inbreeding in cross-fertilizing crops.

They may have various reasons for doing this:

- The development of pure lines (mostly by continued selfing) for use as parents in the breeding of hybrid varieties, *e.g.* in maize or cucumber.
- To promote the efficiency of elimination of an undesired recessive gene
- Maintenance of a genic male sterile 'line'

Note:FS-mating occurs also when a maintaining a genic male sterile barley 'line': male sterile plants are harvested after having been pollinated by their male fertile full sibs. (This is also applied in the case of recurrent selection in self-fertilizing cereals).

Thus the harvesting of a female plant (say genotype *mm*) implies harvest of seed due to the cross *mm* ×*Mm* (where *Mm* represents the genotype assumed for hermaphroditic plants). The genotypic composition of the obtained FS-family is (1/2, 1/2, 0). Repeated application of this procedure implies repeated FSmating.

The most powerful form of inbreeding of cross-fertilizing crops, *e.g.* dioecious crops, occurs with repeated crossing of the type:

- Full sib × full sib, *i.e.* full sib mating, or
- Parent ×× offspring.

Full Sib Mating

The offspring due to a cross of two genotypes constitutes a family. The plants belonging to the family share both their maternal and their paternal parent. With regard to each other these plants are full sibs. Together they form a full sib family (FS-family). Crossing of plants belonging to the same FS-family is called full sib mating (FS-mating). FS-mating may be used when inbreeding of dioecious crops, such as spinach or asparagus, is the aim.

It occurs spontaneously in the case of open pollination within FS-families grown in isolation. This is applied in hermaphroditic, monoecious or dioecious crops in the case of separated FS-family selection. Note describes how FS-mating is applied when maintaining a genic male sterile 'line'.

Parent ×× offspring mating

In this book the notation A ×× B indicates the cross A×B and/or the reciprocal cross B × A. Parent ×× offspring crosses, *i.e.* so-called PO-mating, can only be applied to perennial crops such as oil palm (producing gametes from the age of 4–5 years for many years) or asparagus (with a juvenile phase lasting two years). The parent is still alive when its offspring reach the reproductive phase. Repeated backcrossing implies continued application of crosses of the type 'recurrent parent ×× offspring'.

In the absence of selection the genotype of the offspring becomes identical to the genotype of the recurrent parent (if the recurrent parent has a homozygous genotype) or to the genotypic composition of the possible lines obtained by selfing of the recurrent parent (if the recurrent parent is heterozygous).

In this chapter only loci segregating for not more than two alleles per locus will be considered. For an extensive treatment of the population genetics theory of inbreeding the reader is referred to Allard, Jain and Workman.

DIPLOID CHROMOSOME BEHAVIOUR AND INBREEDING

One Locus with Two Alleles

With continued inbreeding of any (infinitely) large population the genotype frequencies will change from one generation to the other until the frequency of plants with a heterozygous genotype has become zero. Starting from the initial population G0 with genotypic composition (*f*0,0, *f*1,0, *f*2,0), eventually a population with genotypic composition (q, 0, p) will be obtained.

- Starting with some arbitrary genotypic composition

Genotype:

Generation	**aa**	**Aa**	**AA**
S0	f0	f1	f2
S1	$f0+\frac{1}{4}f1$	$\frac{1}{2}f1$	$f2+\frac{1}{4}f1$
S2	$f0+\left(\frac{1}{4}+\frac{1}{8}\right)f1$	$\frac{1}{4}f1$	$f2+\left(\frac{1}{4}+\frac{1}{8}\right)f1$
S3	$f0+\left(\frac{1}{4}+\frac{1}{8}+\frac{1}{16}\right)f1$	$\frac{1}{8}f1$	$f2+\left(\frac{1}{4}+\frac{1}{8}+\frac{1}{16}\right)f1$
.			
.			
S∞	q	0	P

- Starting with F1, *i.e.* a population with genotypic composition (0, 1, 0)

Generation		Inbreeding	Panmictic		Genotype	
(t)	Population	Coefficient	(℘index(P)	aa	Aa	AA
0	S0(= F1)	−1	2	0	1	0
1	S1(= F2)	0	1	$\frac{1}{4}$	$\frac{1}{2}$	$\frac{1}{4}$
2	S2(= F3)	$\frac{1}{2}$	$\frac{1}{2}$	$\frac{3}{8}$	$\frac{2}{8}$	$\frac{3}{8}$
3	S3(= F4)	$\frac{3}{4}$	$\frac{1}{4}$	$\frac{7}{16}$	$\frac{2}{16}$	$\frac{7}{16}$
4	S4(= F5)	$\frac{7}{8}$	$\frac{1}{8}$	$\frac{15}{32}$	$\frac{2}{32}$	$\frac{15}{32}$
5	S5(= F6)	$\frac{15}{16}$	$\frac{1}{16}$	$\frac{31}{64}$	$\frac{2}{64}$	$\frac{31}{64}$
6	S6(= F7)	$\frac{31}{32}$	$\frac{1}{32}$	$\frac{63}{128}$	$\frac{2}{128}$	$\frac{63}{128}$
7	S7(= F8)	$\frac{63}{64}$	$\frac{1}{64}$	$\frac{127}{256}$	$\frac{2}{256}$	$\frac{127}{256}$
∞	S∞(= F∞)	1	0	$\frac{1}{2}$	0	$\frac{1}{2}$

Illustrates this for inbreeding by means of continued selfing. It appears that the genotype frequencies approach, in an asymptotic manner, the gene and haplotype frequencies.

Often the frequency of heterozygous plants in generation *t*, *i.e.* *f*1,*t*, is written in the form,

$$2pq(1-f_t)$$

In this expression the factor 1–*Ft* describes the deviation from the Hardy–Weinberg frequency. The factor is called the panmictic index, sometimes designated by the symbol *P*. This implies that *P* = 1-*Ft*. The parameter *Ft*, say 'script F', is the inbreeding coefficient (or fixation index) pertaining to generation t. When starting with an F1 population, F2 is the first generation due to self-fertilization. For this reason the F2 population is chosen to be generation 1. (Its genotypic composition is equal to the genotypic composition of the population obtained by panmictic reproduction of the F1) Successive generations may be indicated by G1,G2..., but in the case of continued selfing the designations S1, S2, S3... are used as well. A general description of the genotypic composition of any population (inbred or not) is now given by,

	Genotype		
	aa	*Aa*	*AA*
f	$q^2 + pqf_t$	$2pq(1-f_t)$	$P^2 + pqf_t$

In several other books, *e.g.* Falconer and MacKay, the inbreeding coefficient is defined as the probability that the two alleles at any loci of a plant are identical by descent.

This would mean that the inbreeding coefficient of an F2 population obtained from cross *AA* × *aa* is equal to 1 2, because 50% of the plants contain, for locus *A-a,* alleles that are identical by descent (this concerns plants with genotype *aa* or *AA*).

In this book the parameter F is used to quantify the deviations from the Hardy–Weinberg frequencies. In an F2 population such deviations are absent and accordingly its inbreeding coefficient is 0. In Note, it is shown that our definition of the inbreeding coefficient F can be interpreted as the coefficient of correlation of numerical values, *e.g.* gene-effects, assigned to the haplotypes of the uniting gametes.

This is based on the following consideration. With random mating the gene effects of the haplotypes of fusing female and male gametes are independent; in the absence of random mating they are interdependent. With inbreeding they tend to be similar; with outbreeding they tend to be different. Breeding of self-fertilizing crops starts mostly with crossing of homozygous lines. For all loci for which the parental lines have a different homozygous genotype the genotype of the F1 is heterozygous. For these loci $p = q = 1\ 2$ and then the expressions in (3.1) simplify to

	Genotype		
	aa	*Aa*	*AA*
f	$\frac{1}{4}(1+f_t)$	$\frac{1}{2}(1-f_t)$	$\frac{1}{4}(1+f_t)$

As $f1{,}0 = 1\ 2\ (1-F0) = 1$, it follows that $F0 = -1$, *i.e.* a negative value for the inbreeding coefficient. The panmictic index of the F1 amounts for heterozygous loci to $P0 = 2$. In the remainder of this section the decrease in the frequency of heterozygous plants is considered for the three most important regular inbreeding systems, *viz.* self-fertilization, full sib mating and parent × offspring mating. To measure this decrease the parameter λ is defined:

$$\lambda = \frac{2pq(1-f_t)}{2pq(1-f_{t-1})} = \frac{1-f_t}{1-f_{t-1}}$$

This parameter indicates the frequency of heterozygous plants as a proportion of this frequency in the preceding generation. At a smaller value for λ the decrease of *f*1 is stronger.

In the case of selfing the values for λ do not depend on *t*; they are approximately constant when applying full sib mating or parent × offspring. Then λ1 = λ2 = · · · = λ*t*. This implies,

$$f1,t = \lambda f_{1,t-1} = \lambda^2 f1,t-2 = \lambda^t f1,0$$

Self-Fertilization

In the F2 generation, the first generation generated by selfing, the genotype frequencies coincide with the Hardy-Weinberg frequencies.

Thus *f*1,1 = 2*pq*, implying that *F*1, the inbreeding coefficient of F2, is zero. In population F–, approximately obtained after a very large number of generations reproducing by means of selfing, there is complete homozygosity, *i.e.* *f*1,– = 0, implying that *F*–, the inbreeding coefficient of F–, is 1.

The decrease of *f*1, due to continued selfing. That *f*1 is halved by each round of reproduction by means of selfing.

Thus,

$$1 - f_t = \frac{1}{2}(1 - f_{t-1})$$

implying

$$f_t = \frac{1}{2}(1 + f_{t-1})$$

With Regard to Continued selfing the expression:

$$1 - f_t = \frac{1}{2}(1 - f_{t-1})$$

or

$$P_t = \frac{1}{2}P_{t-1}$$

implies

$$P_t = \left(\frac{1}{2}\right)^t P_0 = \left(\frac{1}{2}\right)^{t-1}$$

$$f_t = 1 - \left(\frac{1}{2}\right)^{t-1}$$

At all other systems of inbreeding the reduction of *f*1 is smaller. The minimum value for l is thus attained with selfing. It amounts to l*S* = 1/2.

Full Sib Mating and Parent Offspring Mating

Li showed that for both full sib mating and parent × offspring mating, the relation:

$$f1,t+2=\frac{1}{2}f1,t+1+\frac{1}{4}f1,t$$

Applies. Consider an initial population with genotypic composition (0,1,0), thus $f1,0$ = 1. In this population plants are crossed in pairwise combinations. In the next generation the genotypic composition of the population obtained, which consists of full sib families, is expected to be (1/2, 1/4, 1/2), with $f1,1$ = 1 2.

Continued full sib mating, within the continuously generated FS-families, gives, according to Equation,

$$f1,2=\frac{1}{2}\left(\frac{1}{2}\right)+\frac{1}{4}(1)=\frac{1}{2}, i.e. \lambda_2=1$$

$$f1,3=\frac{1}{2}\left(\frac{1}{2}\right)+\frac{1}{4}\left(\frac{1}{2}\right)=\frac{3}{8}, i.e. \lambda_3=\frac{3}{4}=0.75$$

$$f1,4=\frac{1}{2}\left(\frac{3}{8}\right)+\frac{1}{4}\left(\frac{1}{2}\right)=\frac{5}{16}, i.e. \lambda_4=\frac{5}{6}=0.8333, etc$$

The first round of inbreeding (full sib mating or parent × offspring mating) does not give a decrease of the frequency of heterozygous plants ($\lambda 2$ = 1). Indeed, with full sib mating first FS-families have to be generated. It appears that λ approaches asymptotically the value λFS = λPO = 0.809. As (0.809)3 = 0.53 ≈ 1/2, three generations of reproduction by means of FSmating or parent × offspring mating give the same reduction in $f1$ as a single round of reproduction by selfing.

GENETIC VARIATIONS IN PLANTS

VARIATION

If you have grown Fast Plants® you will have noticed variation among the individual plants within the group. Variation can range from a little to a lot. This chapter is designed to help teachers and students understand the basis of the variation they observe in their plantings. Variation is one of the fundamental characteristics of life. All organisms exhibit some variation among individuals. Understanding the ways that variation is manifested in organisms, how it comes to be expressed through the development of the individual, and how it is transmitted from one individual to the next generation of individuals are central themes in biology.

Working with Fast Plants® will enrich a student's understanding of variation. By observing the growth and development of a Fast Plants® through the various stages in the life cycle, students will become aware of many visible features, or phenotypes, that make up the organism.

Only, however, upon close observation of a population of plants, will they become aware that the characteristics observed on one plant vary more or less on other plants. Such is the nature of variation.

Phenotypic Variation

Phenotypic variation, *e.g.* plant height at a particular stage of development, is considered to be the expression of the genetic makeup (genotype) of the individual as it interacts with the environment. Variation in plant height among individuals in a population is therefore due to variation in the interaction between the genotype and the environment, as expressed through the development of each individual plant.

DESCRIBING AND OBSERVING PHENOTYPIC VARIATION

In order to be useful in an experiment the phenotype must be described using terms that are widely understood and easily communicated. For these reasons scientists have agreed upon various standards or descriptors to describe characteristics in the natural world. Descriptors take many forms, see WFPID Observing and Describing. The choice of how to describe what you observe is important, because it will determine the kinds of descriptors used and establish the basis for recording, analyzing and communicating results.

ENVIRONMENTAL VARIATION

Much can be learned about the role of light, temperature, and nutrients on plant development from experiments in which one or more environmental parameters are changed. Because Fast Plants® are highly responsive to changes in the environment, they are ideal for examining the role of the environment on the expression of phenotypic variation.

Although Fast Plants® are able to grow within a wide range of environmental conditions, for most investigations it is recommended that they be grown under uniform and ideal conditions. In this way variation arising from sub-optimal conditions of environment will be minimized. The Wisconsin Fast Plants® Information Document (WFPID) Understanding the Environment describes how to provide and maintain the various physical, chemical and biotic components of the environment that are most suitable for Fast Plants®.

Genotypic Variation

How can students use Fast Plants® to investigate the contribution of the genotype to the phenotype? Fast Plants®, rapid cycling Brassica rapa, are genotypically variable in that they have a genetically controlled mating system that prevents self-fertilization and favours out-crossing among individuals. As a consequence, even seed stocks selected for uniformity of specific phenotypes and genotypes exhibit considerable variation for other traits.

The Wisconsin Fast Plants® Programme has developed a number of genetic stocks of rapid cycling Brassica rapa for genetic investigations on the nature and inheritance of variation. Some WFP stocks contain distinctive mutant phenotypes, *e.g.*, anthocyaninless plant, anl, yellow green plant, ygr1, rosette, ros, and male sterile, mst2, that are suitable for Mendelian genetics. Other stocks exhibit phenotypes whose expression may vary continually and which may be quantified as discrete or countable units, *e.g.*, number of hairs, or as estimates of size, *e.g.* petite dwarf, dwf1, or of intensity of colour saturation, *e.g.*, purple anthocyanin. These quantitative phenotypes may be conditioned by a few or many genes and normally require numerical descriptions in which the statistical notations of population size, (n), range (r), arithmetic mean (x), and standard deviation (s) are applied.

Yet other stocks have been developed to combine both simply inherited mutant genotypes and quantitatively expressed phenotypes. An important part of WFP is the continuing development and improvement of seed stocks for uses in genetics.

GENETIC DIVERSITY

Genetic diversity is a level of biodiversity that refers to the total number of genetic characteristics in the genetic makeup of a species. It is distinguished from genetic variability, which describes the tendency of genetic characteristics to vary. The academic field of population genetics includes several hypotheses and theories regarding genetic diversity.

The neutral theory of evolution proposes that diversity is the result of the accumulation of neutral substitutions. Diversifying selection is the hypothesis that two subpopulations of a species live in different environments that select for different alleles at a particular locus. This may occur, for instance, if a species has a large range relative to the mobility of individuals within it. Frequency-dependent selection is the hypothesis that as alleles become more common, they become less fit. This is often invoked in host-pathogen interactions, where a high frequency of a defensive allele among the host means that it is more likely that a pathogen will spread if it is able to overcome that allele.

IMPORTANCE OF GENETIC DIVERSITY

There are many different ways to measure genetic diversity. The modern causes for the loss of animal genetic diversity have also been studied and identified. A 2007 study conducted by the National Science Foundation found that genetic diversity and biodiversity are dependent upon each other — that diversity within a species is necessary to maintain diversity among species, and vice versa. According to the lead researcher in the study, Dr. Richard Lankau, "If any one type is removed from the system, the cycle can break down, and the community becomes dominated by a single species."

Survival and Adaptation

Genetic diversity plays a very important role in survival and adaptability of a species because when a species's environment changes, slight gene variations are necessary to produce changes in the organisms' anatomy that enables it to adapt and survive. A species that has a large degree of genetic diversity among its population will have more variations from which to choose the most fit alleles.

Increase in genetic diversity is also essential for an organism to evolve. Species that have very little genetic variation are at a great risk. With very little gene variation within the species, healthy reproduction becomes increasingly difficult, and offspring often deal with similar problems to those of inbreeding. The vulnerability of a population to certain types of diseases can also increase with reduction in genetic diversity.

Agricultural Relevance

When humans initially started farming, they used selective breeding to pass on desirable traits of the crops while omitting the undesirable ones. Selective breeding leads to monocultures: entire farms of nearly genetically identical plants. Little to no genetic diversity makes crops extremely susceptible to widespread disease. Bacteria morph and change constantly. When a disease causing bacterium changes to attack a specific genetic variation, it can easily wipe out vast quantities of the species. If the genetic variation that the bacterium is best at attacking happens to be that which humans have selectively bred to use for harvest, the entire crop will be wiped out.

A very similar occurrence is the cause of the infamous Potato Famine in Ireland. Since new potato plants do not come as a result of reproduction but rather from pieces of the parent plant, no genetic diversity is developed, and the entire crop is essentially a clone of one potato, it is especially susceptible to an epidemic. In the 1840s, much of Ireland's population depended on potatoes for food. They planted namely the "lumper" variety of potato, which was susceptible to a rot-causing plasmodiophorid called *Phytophthora infestans*. This plasmodiophorid destroyed the vast majority of the potato crop, and left tens of thousands of people to starve to death.

Coping with Poor Genetic Diversity

The natural world has several ways of preserving or increasing genetic diversity. Among oceanic plankton, viruses aid in the genetic shifting process.

Ocean viruses, which infect the plankton, carry genes of other organisms in addition their own. When a virus containing the genes of one cell infects another, the genetic makeup of the latter changes. This constant shift of genetic make-up helps to maintain a healthy population of plankton despite complex and unpredictable environmental changes.

Cheetahs are a threatened species. Extremely low genetic diversity and resulting poor sperm quality has made breeding and survivorship difficult for cheetahs -- only about 5% of cheetahs survive to adulthood. About 10,000 years ago, all but the jubatus species of cheetahs died out. The species encountered a population bottleneck and close family relatives were forced to mate with each other, or inbreed. However, it has been recently discovered that female cheetahs can mate with more than one male per litter of cubs. They undergo induced ovulation, which means that a new egg is produced every time a female mates. By mating with multiple males, the mother increases the genetic diversity within a single litter of cubs.

THE TOOLS OF GENETICS: RECOMBINANT DNA AND CLONING

To splice a human gene (in this case, the one for insulin) into a plasmid, scientists take the plasmid out of an *E. coli* bacterium, cut the plasmid with a restriction enzyme, and splice in insulin-making human DNA. The resulting hybrid plasmid can be inserted into another *E. coli* bacterium, where it multiplies along with the bacterium. There, it can produce large quantities of insulin.

In the early 1970s, scientists discovered that they could change an organism's genetic traits by putting genetic material from another organism into its cells. This discovery, which caused quite a stir, paved the way for many extraordinary accomplishments in medical research that have occurred over the past 35 years.

How do scientists move genes from one organism to another? The cutting and pasting gets done with chemical scissors: enzymes, to be specific. Take insulin, for example. Let's say a scientist wants to make large quantities of this protein to treat diabetes. She decides to transfer the human gene for insulin into a bacterium, *Escherichia coli,* or *E. coli,* which is commonly used for genetic research. That's because E. coli reproduces really fast, so after one bacterium gets the human insulin gene, it doesn't take much time to grow millions of bacteria that contain the gene.

The first step is to cut the insulin gene out of a copied, or "cloned," version of the human DNA using a special bacterial enzyme from bacteria called a restriction endonuclease. (The normal role of these enzymes in bacteria is to chew up the DNA of viruses and other invaders.) Each restriction enzyme recognizes and cuts at a different nucleotide sequence, so it's possible to be very precise about DNA cutting by selecting one of several hundred of these enzymes that cuts at the desired sequence. Most restriction endonucleases make slightly staggered incisions, resulting in "sticky ends," out of which one strand protrudes.

The next step in this example is to splice, or paste, the human insulin gene into a circle of bacterial DNA called a plasmid. Attaching the cut ends together is done with a different enzyme (obtained from a virus), called DNA ligase. The sticky ends join back together kind of like jigsaw puzzle pieces. The result: a cut-and-pasted mixture of human and bacterial DNA.

The last step is putting the new, recombinant DNA back into *E. coli* and letting the bacteria reproduce in a petri dish. Now, the scientist has a great tool: a version of *E. coli* that produces lots of human insulin that can be used for treating people with diabetes.

So, what is cloning? Strictly speaking, it's making many copies of a gene—in the example above, *E. coli* is doing the cloning. However, the term cloning is more generally used to refer to the entire process of isolating and manipulating a gene. Dolly the cloned sheep contained the identical genetic material of another sheep. Thus, researchers refer to Dolly as a clone.

RULES GOVERNING RNA'S ANATOMY REVEALED

"RNA is a very floppy molecule that often functions by binding to something else and then radically changing shape," said Al-Hashimi, who is the Robert L. Kuczkowski Professor of Chemistry and a professor of biophysics. These shape changes, in turn, trigger other processes or cascades of events, such as turning specific genes on or off.

Because of the RNA molecule's mercurial nature, "you can't really define it as having a single structure," Al-Hashimi said. "It has many possible orientations, and different orientations are stabilized under different conditions, such as the presence of particular drug molecules."

A major goal in structural biology and biophysics is to be able to predict not only the complex three-dimensional shapes that RNA assumes (which are dictated by the order of its nucleic acid building blocks), but also the various shapes RNA takes on after binding to other molecules such as proteins and small-molecule drugs. Further, researchers would like to be able to manipulate the 3-D structure and resulting activity of RNA by tweaking the drug molecules with which it interacts. But to do that, they need to understand the rules that govern the anatomy of RNA.

The quest has parallels to the study of human anatomy, Al-Hashimi said. "Your body has a specific shape that changes predictably when you are walking or when you are catching a ball; we want to be able to understand these anatomical rules in RNA."

Manipulating RNA is a much sought-after goal, given the recent explosion in vital cellular roles ascribed to RNA and the growing number of diseases that are linked to RNA malfunction. RNA performs many of its roles by serving as a switch that changes shape in response to cellular signals, prompting appropriate reactions in response. The versatile molecule also is essential to retroviruses such as HIV, which have no DNA and instead rely on RNA to both transport and execute genetic instructions for everything the virus needs to invade and hijack its host. In earlier work, Al-Hashimi's team determined that rather than changing shape in response to encounters with drug molecules, RNA goes through a predictable course of shape changes on its own. Drug molecules simply "wait for" the right shape and attach to RNA when the RNA assumes the particular drug's preferred orientation, Al Hashimi said.

But what rules control the predictable path of shapes the RNA molecule assumes? And are those rules the same for all sorts of RNA molecules? In the current work, Al-Hashimi's team investigated those questions.

"RNA is very similar to the human body in its construction, in that it's made up of limbs that are connected at joints," Al-Hashimi said. The limbs are the familiar, ladder-like double helix structures, and the joints are flexible junctions. The prevailing view was that interactions among loopy structures at the tips of the limbs played a role in defining the molecule's overall 3-D shape, much as a handshake defines the orientation of two arms, but Al-Hashimi's group decided to look at things from a different perspective.

"We wondered if the junctions themselves might provide the definition," Al-Hashimi said. "If you look at your arm, you'll notice that you can't move it, relative to your shoulder, in just any way; it's confined to a certain pathway because of the joint's geometry. We wondered if the same thing might be true of RNA."

To investigate that possibility, the researchers turned to a database of RNA structures and found that all structures with two helices linked by a particular type of junction called a trinucleotide bulge fell along the same pathway.

The team then went on to explore structures of RNA molecules with other kinds of junctions. All were confined to similar pathways, but the precise pathway of a given RNA depended upon structural features of its junction. Just as anatomical features of our shoulders, elbows, hips and knees define the range of motion of our arms and legs, the anatomy of RNA's junctions dictates the motion of its helices.

Next, Al-Hashimi and coworkers wanted to understand how drug molecules cause RNA molecules to freeze in specific positions. In earlier work with an RNA molecule known as TAR, which is critical for replication of HIV and thus a key target for anti-HIV drugs, the researchers had found that certain drug molecules froze the RNA molecule in a nearly straight position, while others trapped the molecule in a bent conformation and still others captured positions between the two extremes. But because that project involved a wide variety of drug molecules, it was hard to figure out why certain ones preferred certain orientations.

To explore the issue more methodically, Al-Hashimi's group used a series of aminoglycosides (antibiotics that are known to target RNA) that systematically differed from one another in charge, size and other chemical properties. Size turned out to be the key: bigger aminoglycosides froze RNA in more bent positions; smaller ones favoured straighter RNA structures. Looking more closely, the researchers discovered that the aminoglycoside molecule nestles between two helices and acts like wedge, forcing the helices apart. Examination of other RNA structures bound to small molecules revealed that this rule is not specific to TAR but a more general feature of RNA-small molecule interactions.

"With these findings, it now should be possible to predict gross features of RNA 3-D shapes based only on their secondary structure, which is far easier to determine than is 3-D structure," Al-Hashimi said. "This will make it possible to gain insights into the 3-D shapes of RNA structures that are too large or complicated to be visualized by experimental techniques such as X-ray crystallography and NMR spectroscopy. The anatomical rules also provide a blueprint for rationally manipulating the structure and thus the activity of RNA, using small molecules in drug design efforts and also for engineering RNA sensors that change structure in user-prescribed ways."

9

Plant Transformation

INTRODUCTION

Plant transformation is the introduction of a foreign piece of DNA, confering a specific trait, into host plant tissue. The foreign gene (termed the "transgene") is incorporated into the host plant genome and stably inherited through future generations. The correct regulatory sequences are added to the gene of interest i.e. promoters and terminators, then the DNA is transferred to the plant cell culture using an appropriate vector. The gene is attached to a selectable marker which allows selection for the presence of the transgene. Genes conferring resistance to a specific antibiotic are often used to serve this purpose. Once the plant tissue has been transformed, the cells containing the transgene are selected and regeneration back into whole plants is carried out. This is possible as plant cells are totipotent, which means that they contain all the genetic information to control the development of that cell into a potentially fertile plant. Therefore, the gene is contained in every single plant cell, however, where it is switched on is determined by the promoter which is controlling the gene. Plant transformation can be carried out in a number of different ways depending on the species of plant in question. This is discussed in the sections below. Plant transformation was developed as an alternative to conventional breeding methods which are more laborious than this fairly simple (now routine) laboratory proceedure.

GENETIC TRANSFORMATION

Transformation is the introduction of DNA representing a cloned gene into a cell so that it expresses the protein encoded by the gene. Although the physical insertion of DNA into a cell's nucleus is straightforward, the expression of proteins encoded by that DNA that is not part of a chromosome is often only transient. Introduced DNA that is inserted into one of the chromosomes will be passed during mitosis to all subsequent daughter cells. It is this "stable" transformation that will allow one to introduce one copy of

DNA into one cell, and then allow the one transformed cell to regenerate a complete organism, where each cell contains a copy of that introduced DNA. The manipulation of an organisms DNA by transformation allows unparalleled ability to determine the function of a gene from levels of cell function, to organismal physiology to ecological roles. It also provides a way to dissect the functional significance of parts of the gene or specific amino acid residues of the resulting protein. Transformation additionally allows the engineering of plants or animals to produce novel proteins or specifically remove expression of certain proteins.

There are a number of ways that cloned DNA can be physically introduced into a cell. DNA can be micro-injected into cells, or shot into the cell on the surface of microprojectiles, or enter through holes in the cell membrane induced by a strong electric current. Drosophila and C. elegans are usually transformed through microinjection. Plant transformation can take advantage of a plant pathogenic bacteria (Agrobacterium tumifaciens) that move DNA from a plasmid it carries into plant cells as part of its life cycle. The mechanisms of this movement will be discussed in class. However, it is possible to manipulate the plasmid such that a gene of interest is placed into the plasmid in Agrobacterium so that the bacteria will introduce this DNA into a plant cell.

To transform most plants using Agrobacterium, a single plant cell that has received the new DNA from the bacteria has to be regenerated into a whole plant. This process involves the culturing of the transformed cell to provide replication of that cell. Levels of plant hormones can be manipulated to cause this mass of cells to form roots and shoots of a regenerated plant.This process can take weeks and the details vary from plant to plant. For Arabidospsis, an alternate method has been developed- called dip infiltration. In this method, Agrobacterium carrying the modified plasmid is introduced into the whole plant by submerging the plant in a bacterial solution. Applying a vacuum can help force the bacterial solution into the inner air spaces between plant cells, but this was found to be not necessary. Agrobacterium will move the DNA from its plasmid into many of these cells in the plant. Some of these transformed cells will be used to make the flowers of the plant, including the pollen and ovules. With the self-fertilization possible in Arabidodpis, seeds produced from these flowers will have the introduced gene at a low rate.

The major technical problem of transformation, regardless of the method used, is the low frequency at which it occurs. Only a small fraction of cells where the DNA has entered the nucleus does the DNA get spliced into a chromosome. Thus, one needs a way of identifying those cells or plants that contain the introduce DNA. Usually, one gene included in the introduced DNA is a selectable marker gene - for example a gene that confers resistance against a chemical that kills normal plant cells (antibiotic inbroader sense). Kanamycin is one such antibiotic that kills plant cells. Including a kanamycin resistance gene along with a gene of interest in the Agrobacterium vector allows one to select transformed plants by growing them on kanamycin. Only transformed plants will survive since they express the introduced kanamycin resistance gene. For the Agrobacterium infiltration method, the seed from the infiltrated plants are plated on agar containing kanamycin – the low number of plants containing the introduced DNA will germinate and grow on these plates.

GENETIC ENGINEERING OF PLANTS

Genetic engineering of plants is much easier than that of animals.

There are several reasons for this:

- There is a natural transformation system for plants (the bacterium *Agrobacterium tumefaciens*),
- Plant tissue can redifferentiate (a transformed piece of leaf may be regenerated to a whole plant), and
- Plant transformation and regeneration are relatively easy for a variety of plants.

The soil bacterium *Agrobacterium tumefaciens* ("tumefaciens" meaning tumor-making) can infect wounded plant tissue, transferring a large plasmid, the Ti plasmid, to the plant cell. Part of the Ti (tumor-inducing) plasmid apparently randomly integrates into the chromosome of the plant.

The integrated part of the plasmid contains genes for the synthesis of:

- Food for the bacterium,
- Plant hormones.

Genes from the Ti plasmid that are integrated in the plant chromosome are expressed at high levels in the plant.

Overproduction of the plant hormones leads to continuous growth of the transformed cells, causing plant tumors. Rapid, cancerous growth of the transformed plant tissue obviously is advantageous to the bacterium: more food gets produced.

The Ti plasmid has been genetically modified ("disarmed") by deleting the genes involved in the production of bacterial food and of plant hormones, and inserting a gene that can be used as a selectable marker. Selectable marker genes generally are coding for proteins involved in breakdown of antibiotics, such as kanamycin. Any gene of interest can be inserted into the Ti plasmid as well. In principle, one can thus transform any plant tissue, and select transformants by screening for antibiotic resistance.

However, unfortunately, there are some complications:

- It has proven difficult to transform some monocots (grasses, etc.) by *Agrobacterium*,
- Regeneration of plants from tissue culture or leaf discs is not always possible.

A number of genetically engineered plant varieties have been developed. Traits that have been introduced by transformation include herbicide resistance, increased virus tolerance, or decreased sensitivity to insect or pathogen attack.

Traditionally, most of such genetically engineered plants were tobacco, petunia, or similar species with a relatively limited agricultural application. However, during the past decade it now has become possible to transform major staples such as corn and rice and to regenerate them to a fertile plant. Increasingly, the transformation procedures used do not depend on *Agrobacterium tumefaciens*. Instead, DNA can be delivered into the cells by small, μm-sized tungsten or gold bullets coated with the DNA.

The bullets are fired from a device that works similar to a shotgun. The modernized device uses a sudden change in pressure of He gas to propel the particles, but the principle of "shooting" the DNA into the cell remains the same.

This DNA-delivery device is nicknamed "gene gun", and has been shown to work for DNA delivery into chloroplasts as well. Over the last several years, use of the "gene gun" has become a very common method to transform plants, and has been shown to be applicable to virtually all species investigated. For example, transformation of rice by this method is now routine. This is a very important development as rice is the most important crop in the world in terms of the number of people critically dependent on it for a major part of their diet.

Another method to get foreign genes into cereals is by electroporation: a jolt of electricity is used to puncture self-repairing holes in protoplasts (*i.e.*, the cell without the cell wall), and DNA can get in through these holes. However, it is often very difficult to regenerate fertile plants from protoplasts of cereals.

Nonetheless, significant advances in overcoming these practical difficulties have been made over the years. Now even transgenic trees have been created: for example, the gene for a coat protein of the plum pox virus has been introduced into apricot. The plum pox virus leads to the feared Sharka disease, for which there is no cure. The resulting transgenic tree shows a markedly decreased sensitivity to this virus. The reason why continuous exposure of the tree to the viral coat protein leads to tolerance against viral infection is not yet understood, however.

Thus, now there are a number of different techniques to introduce foreign genes into plants. Essentially all major crop plants can be (and have been or are being) genetically engineered, the procedures are now routine and the frequency of success is very high. Even though genetically engineered crops are more costly than the usual ones, US farmers have rather readily accepted them provided that tangible benefits can be demonstrated. However, it is questionable whether the farmer in poorer countries can come up with the funds to "try out" and use the new crops. Another issue in this respect is how genetically engineered crops are perceived by the consumer.

Even though in the US there is little resistance to such crops as long as the products can be shown to be safe and advantageous, in other countries genetically modified foods are received poorly by the consumer. It is unlikely that there is a rationally sound basis for this rather hostile reaction of the consumer, as most of the crops are the results of human manipulation (such as centuries of breeding) and may have been treated with harmful herbicides and pesticides.

Time and education will need to be invested to provide consumers and consumer advocates with a balanced opinion on the acceptability of the origin of their foods. On the other hand, as so many plants (soybean, corn, etc.) are genetically modified and the nature of the genetic modification is not necessarily easy to explain, it may be simpler to label those foods that are guaranteed free of "genetically modified organisms" or their products. However, keep in mind that traditional breeding has genetically modified essentially all agricultural products, so it may be difficult to define what is actually free of genetically modified organisms.

TRANSGENIC CROPS

H- Transgenic crops: The varieties created through the use of a biotechnological approach commonly known as recombinant DNA, genetic engineering (GE), transgenic modification, or genetic modification (GM). The products of genetic engineering are often called genetically modified organisms, or GMOs. All these terms refer to methods of recombinant DNA technology by which biologists splice genes from one or more species into the DNA of crop plants to transfer chosen genetic traits. Genes are segments of DNA that controlled the information that determines the structure and function of a living organism. Genetic engineers are molecular biologists who manipulate this information, typically by taking genes from one species—an animal, plant, bacterium, or virus—and inserting them into another species, such as an agricultural crop.

With the advent of genetic engineering of plants around 1983, it appeared that this new biotechnology would benefit and even revolutionize agriculture. The transfer of desirable genetic traits across species barriers has shown promise for solving problems in the management of agricultural crops. Potential benefits include reduced toxic pesticide use, improved weed control resulting in less tillage and soil erosion, and water conservation.

ECOLOGICAL ISSUES

Gene flow to neighbouring crops and to related wild species. Ecological scientists have little doubt that gene flow from transgenic fields into conventional crops and related wild plants will occur. Gene flow from transgenic to conventional crops is of concern to farmers because of its potential to cause herbicide resistance in related conventional crops. For example, in western Canada, three different herbicide-resistant canola varieties have cross-pollinated to create canola plants that are resistant to all three types of herbicide.

Through what is called gene-stacking, this new triple resistance has turned volunteer canola into a significant weed problem. Gene flow from transgenic crops to wild relatives creates a potential for wild plants or weeds to acquire traits that improve their fitness, turning them into "super weeds." For example, if jointed goatgrass—a weedy relative of wheat—acquires the herbicide-tolerant trait of Roundup Ready wheat, it will thrive in crop fields unless applications of other herbicides are made. There is already evidence of such outcrossing from herbicide-resistant wheat to jointed goatgrass. Frank Young and his colleagues at Washington State University found that imidazolone-resistant wheat (not a transgenic variety) outcrossed to goatgrass in one season. Other traits that wild plants could acquire from transgenic plants that would increase their weediness are insect and virus resistance. Because of their experience with classically bred plants, few scientists doubt that genes will move from crops into the wild: seven of the world's thirteen most important crop weeds have been made weedier by genes acquired from classically bred crops. Because gene flow has the potential to affect farmers' crop and pest management, crop marketability, and liability, more research needs to be done to determine the conditions under which gene flow from transgenic plants is likely to be significant.

Pesticide Resistance in Insect Pests

Bt has been widely used as a microbial spray because it is toxic only to caterpillars. In fact, it is a pest management tool that organic farmers depend on—one of the few insecticides acceptable under organic rules. Unlike the commercial insecticide spray, the Bt engineered into crop plants is reproduced in all, or nearly all, the cells of every plant, not just applied on the plant surface for a temporary toxic effect. As a result, the possibility that transgenic Bt crops will accelerate insect pests' development of resistance to Bt is a serious concern. Pest resistance to Bt would remove this valuable and environmentally benign tool from farmers' and forest managers' pest control toolbox. Antibiotic resistance The use of antibiotic-resistant marker genes for the delivery of a gene package into a recipient plant carries the danger of spreading antibiotic-resistant bacteria. The likely result will be human and animal health diseases resistant to treatment with available antibiotics. Research is needed on antibiotic resistance management in transgenic crops. Already the European Commission's new rules governing transgenic crops stipulate phasing out antibiotic-resistant marker genes by the end of 2004.

Effects on Beneficial Organisms

Evidence is increasing that transgenic crops—either directly or through practices linked to their production—are detrimental to beneficial organisms. New studies are finding that Bt crops exude Bt in concentrations high enough to be toxic to some beneficial soil organisms. In addition, a study by University of Arkansas scientists has shown that "root development, nodulation and nitrogen fixation are impaired" in some varieties of Roundup Ready soybeans. The reason is that the beneficial rhizobium responsible for nitrogen fixation in soybeans is sensitive to Roundup. It also appears that disruption of beneficial soil organisms can interfere with plant uptake of phosphorus, an essential plant nutrient.

Insecticidal crops in two ways can affect beneficial insects that prey on insect pests. First, the Bt in transgenic insecticidal crops has been shown in some laboratory studies to be toxic to ladybird beetles, lacewings, and monarch butterflies. The extent to which these beneficials are affected in the field is a matter of further study. Second, because the insecticidal properties of Bt crops function even in the absence of an economic threshold of pests, Bt crops potentially can reduce pest populations to the point that predator species are negatively affected.

Reduced Crop Genetic Diversity

As fewer and larger firms dominate the rapidly merging seed and biotechnology market, transgenic crops may continue the trend towards simplification of cropping systems by reducing the number and type of crops planted. In addition, seed saving, which promotes genetic diversity, is restricted for transgenic crops.

Food Safety

Because food safety is predominantly a concern of consumers, However, because food safety issues do impact marketing and international trade, here is a short list of consumer safety concerns about transgenic food:

- Possibility of toxins in food
- Possibility of new pathogens
- Reduced nutritional value
- Introduction of human allergens
- Transfer of antibiotic resistance to humans
- Unexpected immune-system and genetic effects from the introduction of novel compounds

Farm Management Issues

The most widely planted transgenic crops on the market today can simplify short-term pest management for farmers and ranchers. In the case of herbicide-tolerant crops, ideally farmers can use a single broad-spectrum herbicide for all their crop weeds, though they may need more than one application in a season. By planting insecticidal crops, farmers can eliminate the need to apply pesticides for caterpillar pests like the European corn borer or the cotton bollworm, though they still have to contend with other crop pests.

While these crops offer handy pest control features, they may complicate other areas of farm management. Farmers who are growing both transgenic and conventional varieties of the same crop will need to segregate the two during all production, harvesting, storage, and transportation phases if they are selling into differentiated markets or plan to save their own seed from their conventional crops.

To minimize the risk of gene flow from transgenic to adjacent conventional crop fields, federal regulations require buffer strips of conventional varieties around transgenic fields. Different transgenic crops require different buffer widths. Because the buffer strips must be managed conventionally, producers have to be willing to maintain two different farming systems on their transgenic fields.

Crops harvested from the buffer strips must be handled and marketed as though they are transgenic. Planted refuges—where pest species can live outside fields of insecticidal and herbicide-tolerant transgenic crops—are also required to slow the development of weed and insect pest resistance to Bt and broad-spectrum herbicides. These refuges allow some individuals in the pest population to survive and carry on the traits of pesticide susceptibility. Requirements governing the size of refuges differ according to the type of transgenic crop grown.

Farmers growing herbicide-tolerant crops need to be aware that volunteer crop plants the following year will be herbicide-resistant. This herbicide resistance makes no-till or direct-seed systems difficult because volunteers can't be controlled with the same herbicide used on the rest of the crop. In a no-till system that relies on the same broad-spectrum herbicide that the volunteer plants are resistant to, these plants will contaminate the harvest of a following conventional variety of the same crop, a situation farmers need to avoid for two reasons.

First, the contamination means a following conventional crop will have to be sold on the transgenic market. This leads to the second reason. If farmers grow and market a transgenic crop for which they do not have a technology agreement and did not pay royalty fees, they face possible prosecution by the company that owns the transgenic variety.

Many farmers have been charged with "theft" of a company's patented seed as a result of contamination in the field Farmers growing insecticidal crops need to recognize that insect pressure is difficult to predict and may not warrant the planting of an insecticidal variety every year. In a year when pest pressure is low, the transgenic seed becomes expensive insurance against the threat of insect damage. Farmers growing transgenic crops need to communicate with their neighbours to avoid contaminating their fields and to ensure that buffers are adequate. In Maine, farmers growing transgenic crops are now required by law to be listed with the state agriculture department, to help identify possible sources of cross-contamination when it occurs. The new law, L.D. 1266, also "requires manufacturers or seed dealers of genetically engineered plants, plant parts or seeds to provide written instructions to all growers on how to plant, grow, and harvest the crops to minimize potential cross-contamination of non-genetically engineered crops or wild plant populations". Farm management issues common to all transgenic crops include yield, cost, price, profitability, management flexibility, sustainability, market acceptance, and liability.

Crop Yield, Costs, and Profitability

Some farmers will get higher yields with a particular transgenic crop variety than with their conventional varieties, and some will get lower yields. The same is true for costs. Some producers will see their overall profitability rise, and some will see it drop.

If the results of research studies so far on yield, chemical use, costs, crop prices, and profitability of these crops seem contradictory, it is because farmers and farms are not all alike. Neither are the conventional varieties used for comparison. Some yield, cost, and profitability trends do appear to be emerging from the growing body of research data for transgenic crops, however. As noted in the Wallace Centre report, Roundup Ready soybeans were designed simply to resist a particular chemical herbicide, not to increase yields. In contrast, Bt corn and cotton, by resisting insect pests, may result in higher yields from reduced pest pressure.

Yield: Herbicide-tolerant Crops—soybeans, Cotton, Canola

Herbicide-tolerant soybeans appear to suffer what's referred to as a "yield drag." Again, in some areas and on some farms this tendency of Roundup Ready soybean varieties to yield less than their comparable, conventional counterparts varies, but overall, they appear to average yields that are five to ten per cent lower per acre.. Drought conditions worsen the effects. The bacterium that facilitates nodulation and nitrogen fixation in the root zone apparently is sensitive to both Roundup and drought.

Yield: Insecticidal Crops—corn, Cotton

Bt cotton, especially, outpaces yields of conventional cotton by as much as 9 to 26 per cent in some cases, though not at all in others. Yield increases for Bt corn have not been as dramatic. Time will tell whether farmers can expect yield increases or decreases in the long run with these and other transgenic crop varieties.

Changes in Chemical Pesticide use

One of the promises of biotechnology is that it will reduce pesticide use and thereby provide environmental benefits and reduce farmers' costs. The herbicide-tolerant and insecticidal varieties are designed specifically to meet these goals.

Increased future pesticide use resulting from the buildup of resistance to heavily used herbicides is a long-term concern. Pesticide use depends on the crop and its specific traits; weather; severity of pest infestations; farm management; geographic location of the farm; and other variables. As a result, conclusions drawn by various studies analysing pesticide use on transgenic crops remain controversial. According to the Wallace Centre report, in a review of the data available up through 2000, crops engineered to contain Bt appear to have decreased the overall use of insecticides slightly, while the use of herbicide-resistant crops has resulted in variable changes in overall herbicide use, with increases in use of some herbicides in some places and decreases in others

Profitability

Farmers need to consider all the factors that determine profitability. No single factor can tell the whole story. Transgenic crop seeds tend to be more costly, and farmers have the added expense of a substantial per-acre fee charged by the owners of transgenic varieties. These costs have to be considered along with input cost changes—whether herbicide or insecticide use and costs go down, go up, or stay the same.

Market price is another factor: Prices for some transgenic crops in some markets are lower than prices for comparable conventional crops, though rarely are they higher. Farmers need to watch the markets. Some buyers will pay a premium for a non-transgenic product, though as transgenic seeds find their way into conventional transportation, storage, and processing streams, these premiums may disappear along with confidence that "GMO-free" products are in fact truly free of engineered genes.

Marketing and Trade

Buyer acceptance is a significant marketing issue for farmers raising transgenic crops. Farmers need to know before they plant what their particular markets will or won't accept. Since most grain handlers cannot effectively segregate transgenic from non-transgenic crops in the same facility, many companies are channeling transgenic crops into particular warehouses.

Farmers need to know which ones and how far away those are. While the European Union officially lifted its two-year moratorium on the introduction of new transgenic crops in February 2001, during the debate over labeling and traceability regulations the moratorium remains in effect. Under the proposed new EU requirements, "all foods and animal feed derived from

GMOs have to be labeled and, in the case of processed goods, records have to be kept throughout the production chain allowing the GMO to be traced back to the farm".

Liability

The need to separate transgenic crops from conventional and organic ones opens farmers to liability for their product at every step from seed to table.

Effective systems for segregation do not exist at present, and will be costly to develop and put into place. Farmers may well end up bearing the added costs of crop segregation, traceability, and labeling.

In the meantime, farmers who grow transgenic varieties—and, ironically, those who do not—are liable for transgenic seeds ending up where they aren't wanted: in their own non-transgenic crop fields, in neighbours' fields, in truckloads of grain arriving at the elevator, in processed food products on retail shelves, and in ships headed overseas.

Farmers who choose not to grow transgenic varieties risk finding transgenic plants in their fields anyway, as a result of cross-pollination via wind, insects, and birds, which may bring pollen from transgenic crops planted miles away.

Besides pollen, sources of contamination include contaminated seed and seed brought in by passing trucks or wildlife. Those farmers whose conventional or organic crops are contaminated, regardless of the route, risk lawsuits filed against them by the companies that own the proprietary rights to seed the farmer didn't buy.

Likewise, farmers who grow transgenic crops risk being sued by neighbours and buyers whose non-transgenic crops become contaminated.

Because contamination by transgenic material has become so prevalent in such a short time, all farmers in areas of transgenic crop production are at risk. Insurance, the most common recourse for minimizing potential losses because of liability, is not available to the nation's farmers for this risk because insurance companies do not have enough information to gauge the potential losses.

A full risk assessment and legal clarification of the distribution of liability among farmers, seed companies, grain handlers, processors, and retailers is needed before farmers can rest assured that transgenic crops won't result in lawsuits against them.

For farmers who have adopted transgenic technology—and for those who have not—in the words of Brian Leahy, executive director of California Certified Organic Farmers, "This technology does not respect property rights". In essence, farmers who grow transgenic crops on some of their fields, and farmers who grow none, risk bearing tremendous liability. This situation won't change until farmers either gain legal protection or until they all grow transgenic crops on all their fields.

Influences on Public Research

While transgenic crop varieties are generally the property of private corporations, those corporations often contract with public-sector agricultural research institutions for some of their development work. In fact, private investment in agricultural research, including germplasm development, has surpassed public investment in recent years. With this shift in funding sources, the following questions become important: Is the private sector unduly influencing the public research agenda? Are corporations directing public research in less socially valuable directions while research on, for instance, sustainable agriculture goes wanting? Are the outcomes of corporate-funded transgenic research and development by our public institutions equitable across the food and agricultural sectors? Is equity even a consideration of our public institutions when they accept this work? Because intellectual property rights (patents) apply to living organisms, making them private property, the free flow of scientific information that has historically characterized public agricultural research is being inhibited. What are the implications for the future of agriculture and society of the secrecy that now surrounds so much of what was formerly shared public knowledge? For a brief history of intellectual property rights as they apply to living organisms.

These and other questions need to be addressed by citizens and their public institutions. These issues are of particular concern to farmers and consumers who would benefit from research into alternative technologies that are less costly (in every way), less risky, and more equitable. Equity requires that the economic benefits and risks of transgenic technology be fairly distributed among technology providers, farmers, merchants, and consumers.

For long-term sustainability, farmers need research that focuses on farms as systems, with internal elements whose relationships can be adjusted to achieve farm management goals. In contrast, transgenic crop research so far has focused on products that complement toxic chemical approaches to control of individual pest species. These are products that can be commercialized by large agribusiness or agri-chemical interests and that farmers must purchase every year. This research orientation only perpetuates the cost-price squeeze that continues to drive so many out of farming.

Industry Concentration and Farmers' Right to Save Seed

The broadening of intellectual property rights in 1980 to cover living organisms, including genes, has resulted in a flurry of mergers and acquisitions in the seed and biotech industries. According to the Wallace Centre report, "Relatively few firms control the vast majority of commercial transgenic crop technologies. These firms have strategically developed linkages among the biotechnology, seed, and agri-chemical sectors to capture as much market value as possible. However, these tightly controlled linkages of product sectors raise serious issues of market access, product innovation, and the flow of public benefits from transgenic crops. Unlike Plant Variety Protection—which does not allow for the patenting of individual genes, but only of crop varieties—Intellectual Property Rights prohibit farmers from saving their seed and undertaking their own breeding programme, and prohibit plant breeders from using the material to create a new generation of varieties adapted to specific regions or growing conditions. Farmers can't save proprietary seed for planting and so must purchase new seed every year. In addition, farmers choosing transgenic varieties must sign a contract with the owner of the variety and pay a substantial per-acre technology fee, or royalty.

The anticipated commercial introduction of transgenic wheat represents a dramatic shift in an industry in which farmers still widely save their own seed. As non-transgenic varieties become contaminated with transgenic ones, even those farmers who choose to stick with conventional varieties will lose the right to replant their own seed. This loss has already occurred in Canada's canola industry, with Monsanto winning its court case against farmer Percy Schmeiser for replanting his own canola variety that had become contaminated with Monsanto's Roundup Ready canola.

The adoption of transgenic crop varieties has brought with it an increasing prevalence of contract production. While contract production can lead to increased value and reduced risk for growers, farmers are justified in their concern about their loss of control when they sign a contract with a private company. Issues associated with contract production of transgenic crops must be considered within the broader context of a sustainable agriculture to include ownership, control, and social equity.

GENETIC TRANSFORMATION IN PLANTS

AGROBACTERIUM MEDIATED GENE TRANSFER

It is the most widely used transformation technique in plants. Agrobacterium tumefaciens, a soil bacteria, contains a Ti plasmid (tumor-inducing) which normally infects dicotyledonous plant cells, making the bacteria an excellent vector for the transfer of foreign DNA. By removing the tumor inducing genes and replacing them with the genes of interest, efficient transformation can occur.

A.tumefaciens naturally infects the wound sites in dicotyledonous plant causing the formation of the crown gall tumors. Later it was evident that A. tumefaciens is capable to transfer a particular DNA segment (T-DNA) of the tumor-inducing (Ti) plasmid into the nucleus of infected cells where it is subsequently stable integrated into the host genome and transcribed.

The T-DNA contains two types of genes:

1. Oncogenic genes, encoding for enzymes involved in the synthesis of auxins and cytokines and responsible for tumor formation;

2. Genes encoding for the synthesis of opines, a product resulted from condensation between amino acids and sugars, which are produced and excreted by the crown gall cells and consume by A. tumefaciens as carbon and nitrogen sources.

Outside the T-DNA, are located the genes for the opine catabolism. These genes are involved in the process of T-DNA transfer from the bacterium to the plant cell and for the bacterium-bacterium plasmid conjugative transfer genes. Adaptation of the bacterial model for foreign gene transfer to crop plant requires the following steps.

- Introduction of the foreign gene into the T-DNA of the bacteria
- Introduction of the bacteria containing the foreign gene into cells of host plant.
- Integrartion of the foreign gene into the genome of the host cell.
- Expression of the foreign gene in the regenerated crop plant
- Transmission of the foreign gene and its expression through normal sexual processes to plants in succeeding generation in seed reproduced species, or through normal sexual propagation in vegetatively reproduced species.

Agrobacterium mediated transformation is the easiest and most simple plant transformation. Plant tissue (often leaves) is cut into small pieces, eg. 10x10mm, and soaked for 10 minutes in a fluid containing suspended *Agrobacterium*. Some cells along the cut will be transformed by the bacterium that inserts its DNA into the cell. Placed on selectable rooting and shooting media, the plants will regrow. Some plants species can be transformed just by dipping the flowers into suspension of *Agrobacterium* and then planting the seeds in a selective medium.

10

Plant Reproduction System

INTRODUCTION

Plants have reproductive systems that are more variable than those in most animals. In many plants, both sexual and asexual means of reproduction are possible. The combination of both means of reproduction can be highly advantageous for plants, especially for perennials. When sexual reproduction is prevented or averted, either by physical damage to the plants or by unfavorable environmental conditions, perennials usually can persist vegetatively until conditions are favorable for sexual reproduction. The main advantage of sexual reproduction is that it provides an opportunity for organisms to produce new combinations of genes or to pass on to succeeding generations new mutations that might be highly adaptive to changing environments or to other conditions. Recombination occurs through the reassortment of parental genes during meiosis and subsequent combination of those genes during fertilization.

The plant kingdom is divided into many classes and divisions. In 1883, Eichler separated the plant kingdom (Plantae) into two subkingdoms, the first being Cryptogamae (bacteria, algae, fungi, molds, liverworts, moss and ferns); the second being Phanerogamae (including gymnosperms and angiosperms). There have been many changes to the classification system, as we learn more about the genetics of different species. More recent taxonomy created new kingdoms that separated fungi, algae, and molds from what we commonly consider as plants. Plantae was separated into two divisions: Bryophytes (non-vascular) and Tracheophytes (vascular). Of the vascular plants, angiosperms (the flowering plants) make up approximately 300,000 of the 312,000 known number of living species. The number of 312,000 is probably a gross under-estimation. For anyone interested there is an estimated 20,000 species of non-vascular plants. There are many different methods of reproduction that plants have developed over the centuries. If a plant's method of reproduction makes it easier to cross species or hybridized, it is more likely that its offspring will be better adapted to a changing environment For more than 8,000 years, human beings have also been affecting plant development by purposefully breeding new characteristics into plants, creating new species. The most important fact regarding reproduction is that every living organism has, in its cell's nucleus, a set of chromosomes. Chromosomes carry genes that have all the information necessary to make new individual organisms.

DIAGRAM OF A DICOT PLANT

One of the simplest methods of reproduction is when one cell divides and becomes two equal halves that will grow large enough to split again. With plants and animals that reproduce this way, each generation is identical to the one before.

CELL DIVIDING DIAGRAM WITH ONE PARENT CELL

Plant reproduction may be sexual, in which two parents produce a genetically different individual; or asexual, involving the propagation of plants that are genetically identical to the parent.

REPRODUCTION IN PLANTS

Plant reproduction is the process by which plants generate new individuals, or offspring. Reproduction is either sexual or asexual. Sexual reproduction is the formation of offspring by the fusion of gametes. Asexual reproduction is the formation of offspring without the fusion of gametes. Sexual reproduction results in offspring genetically different from the parents. Asexual offspring are genetically identical except for mutation. In higher plants, offspring are packaged in a protective seed, which can be long lived and can disperse the offspring some distance from the parents. In flowering plants (angiosperms), the seed itself is contained inside a fruit, which may protect the developing seeds and aid in their dispersal.

SEXUAL REPRODUCTION IN ANGIOSPERMS: OVULE FORMATION

All plants have a life cycle that consists of two distinct forms that differ in size and the number of chromosomes per cell. In flowering plants, the large, familiar form that consists of roots, shoots, leaves, and reproductive structures (flowers and fruit) is diploid and is called the sporophyte. The sporophyte produces haploid microscopic gametophytes that are dependent on tissues produced by the flower. The reproductive cycle of a flowering plant is the regular, usually seasonal, cycling back and forth from sporophyte to gametophyte.

The flower produces two kinds of gametophytes, male and female. The female gametophyte arises from a cell within the ovule, a small structure within the ovary of the flower. The ovary is a larger structure within the flower that contains and protects usually many ovules. Flowering plants are unique in that their ovules are entirely enclosed in the ovary. The ovary itself is part of a larger structure called the carpel, which consists of the stigma, style, and ovary. Each ovule is attached to ovary tissue by a stalk called the funicle. The point of attachment of the funicle to the ovary is called the placenta.

As the flower develops from a bud, a cell within an ovule called the archespore enlarges to form an embryo-sac mother cell (EMC). The EMC divides by meiosis to produce four megaspores. In this process the number of chromosomes is reduced from two sets in the EMC to one set in the megaspores, making the megaspores haploid. Three of the four megaspores degenerate and disappear, while the fourth divides mitotically three times to produce eight haploid cells. These cells together constitute the female gametophyte, called the embryo sac.

The eight embryo sac cells differentiate into two synergids, three antipodal cells, two fused endosperm nuclei, and an egg cell. The mature embryo sac is situated at the outer opening (micropyle) of the ovule, ready to receive the sperm cells delivered by the male gametophyte.

Pollen

The male gametophyte is the mature pollen grain. Pollen is produced in the anthers, which are attached at the distal end of filaments. The filament and anther together constitute the stamen, the male sex organ. Flowers usually produce many stamens just inside of the petals. As the flower matures, cells in the anther divide mitotically to produce pollen mother cells (PMC). The PMCs divide by meiosis to produce haploid microspores in groups of four called tetrads. The microspores are housed within a single layer of cells called the tapetum, which provides nutrition to the developing pollen grains.

Each microspore develops a hard, opaque outer layer called the exine, which is constructed from a lipoprotein called sporopollenin. The exine has characteristic pores, ridges, or projections that can often be used to identify a species, even in fossil pollen. The microspore divides mitotically once or twice to produce two or three haploid nuclei inside the mature pollen grain. Two of the nuclei function as sperm nuclei that can eventually fuse with the egg and endosperm nuclei of the embryo sac, producing an embryo and endosperm, respectively.

For sexual fusion to take place, however, the pollen grain must be transported to the stigma, which is a receptive platform on the top of the style, an elongated extension on top of the carpel(s). Here the moist surface or chemicals cause the pollen grain to germinate. Germination is the growth of a tube from the surface of a pollen grain. The tube is a sheath of pectin, inside of which is a solution of water, solutes, and the two or three nuclei, which lack any cell walls. Proper growth of the pollen tube requires an aqueous solution of appropriate solute concentration, as well as nutrients such as boron, which may aid in its synthesis of pectin.

At the apex of the tube are active ribosomes and endoplasmic reticulum (types of cell organelles) involved in protein synthesis. Pectinase and a glucanase (both enzymes that break down carbohydrates) probably maintain flexibility of the growing tube and aid in penetration. The pollen tube apex also releases ribonucleic acid (RNA) and ribosomes into the tissues of the style. The tube grows to eventually reach the ovary, where it may travel along intercellular spaces until it reaches a placenta. Through chemical recognition, the pollen tube changes its direction of growth and penetrates through the placenta to the ovule. Here the tube reaches the embryo sac lying close to the micropyle, and sexual fertilization takes place.

Double Fertilization

Fertilization in flowering plants is unique among all known organisms, in that not one but two cells are fertilized, in a process called double fertilization. One sperm nucleus in the pollen tube fuses with the egg cell in the embryo sac, and the other sperm nucleus fuses with the diploid endosperm nucleus. The fertilized egg cell is a zygote that develops into the diploid embryo of the sporophyte. The fertilized endosperm nucleus develops into the triploid endosperm, a nutritive tissue that sustains the embryo and seedling. The only other known plant group exhibiting double fertilization is the Gnetales in the genus *Ephedra,* a nonflowering seed plant. However, in this case the second fertilization product degenerates and does not develop into endosperm.

Double fertilization begins when the pollen tube grows into one of the two synergid cells in the embryo sac, possibly as a result of chemical attraction to calcium. After penetrating the synergid, the apex of the pollen tube breaks open, releasing the two sperm nuclei and other contents into the synergid. As the synergid degenerates, it envelops the egg and endosperm cells, holding the two sperm nuclei close and the other expelled contents of the pollen tube. The egg cell then opens and engulfs the sperm cell, whose membrane breaks apart and allows the nucleus to move near the egg nucleus. The nuclear envelopes then disintegrate, and the two nuclei combine to form the single diploid nucleus of the zygote. The other sperm cell fuses with the two endosperm nuclei, forming a single triploid cell, the primary endosperm cell, which divides mitotically into the endosperm tissue.

Double fertilization and the production of endosperm may have contributed to the great ecological success of flowering plants by accelerating the growth of seedlings and improving survival at this vulnerable stage. Faster seedling development may have given flowering plants the upper hand in competition with gymnosperm seedlings in some habitats, leading to the abundance of flowering plants in most temperate and tropical regions.

Gymnosperms nevertheless are still dominant at higher elevations and latitudes, and at low elevations in the Pacific Northwest coniferous forests, such as the coastal redwoods. The reasons for these patterns are still controversial.

The Seed

The seed is the mature, fertilized ovule. After fertilization, the haploid cells of the embryo sac disintegrate. The maternally derived diploid cells of the ovule develop into the hard, water-resistant outer covering of the seed, called the testa, or seed coat. The diploid zygote develops into the embryo, and the triploid endosperm cells multiply and provide nutrition. The testa usually shows a scar called the hilum where the ovule was originally attached to the funicle. In some seeds a ridge along the testa called the raphe shows where the funicle originally was pressed against the ovule. The micropyle of the ovule usually survives as a small pore in the seed coat that allows passage of water during germination of the seed. In some species, the funicle develops into a larger structure on the seed called an aril, which is often brightly colored, juicy, and contains sugars that are consumed by animals that may also disperse the seed (as in nutmeg, arrowroot, oxalis, and castor bean). This is distinct from the fruit, which forms from the ovary itself.

The embryo consists of the cotyledon(s), epicotyl, and hypocotyl. The cotyledons resemble small leaves, and are usually the first photosynthetic organs of the plant. The portion of the embryo above the cotyledons is the epicotyl, and the portion below is the hypocotyl. The epicotyl is an apical meristem that produces the shoot of the growing plant and the first true leaves after germination.

The hypocotyl develops into the root. Often the tip of the hypocotyl, the radicle, is the first indication of germination as it breaks out of the seed. Flowering plants are classified as monocotyledons or dicotyledons (most are now called eudicots) based on the number of cotyledons produced in the embryo. Common monocotyledons include grasses, sedges, lilies, irises, and orchids; common dicotyledons include sunflowers, roses, legumes, snapdragons, and all nonconiferous trees.

The endosperm may be consumed by the embryo, as in many legumes, which use the cotyledons as a food source during germination. In other species the endosperm persists until germination, when it is used as a food reserve. In grains such as corn and wheat, the outer layer of the endosperm consists of thick-walled cells called aleurone, which are high in protein.

The Fruit

The fruit of a flowering plant is the mature ovary. As seeds mature, the surrounding ovary wall forms a protective structure that may aid in dispersal. The surrounding ovary tissue is called the pericarp and consists of three layers.

From the outside to inside, these layers are the exocarp, mesocarp, and endocarp. The exocarp is usually tough and skinlike. The mesocarp is often thick, succulent, and sweet. The endocarp, which surrounds the seeds, may be hard and stony, as in most species with fleshy fruit, such as apricots.

A fruit is termed simple if it is produced by a single ripened ovary in a single flower (apples, oranges, apricots). An aggregated fruit is a cluster of mature ovaries produced by a single flower (blackberries, raspberries, strawberries). A multiple fruit is a cluster of many ripened ovaries on separate flowers growing together in the same inflorescence (pineapple, mulberry. A simple fruit may be fleshy or dry. A fleshy simple fruit is classified as a berry (grape, tomato, papaya), pepo (cucumber, watermelon, pumpkin), hesperidium (orange), drupe (apricot), or pome (apple).

Dry simple fruits have a dry pericarp at maturity. They may or may not dehisce, or split, along a seam to release the seeds. A dehiscent dry fruit is classified as legume or pod (pea, bean), silique or silicle (mustard), capsule (poppy, lily), or follicle (milkweed, larkspur, columbine). An indehiscent dry fruit that does not split to release seeds is classified as an achene (sunflower, buttercup, sycamore), grain or caryopsis (grasses such as corn, wheat, rice, barley), schizocarp (carrot, celery, fennel), winged samara (maple, ash, elm), nut (acorn, chestnut, hazelnut), or utricle (duckweed family). Some fruiting bodies contain non-ovary tissue and are sometimes called pseudocarps. The sweet flesh of apples and pears, for example, is composed not of the pericarp but the receptacle, or upper portion, of the flowering shoot to which petals and other floral organs are attached.

Fruiting bodies of all kinds function to protect and disperse the seeds they contain. Protection can be physical (hard coverings) or chemical (repellents of seed predators). Sweet, fleshy fruits are attractive food for birds and mammals that consume seeds along with the fruit and pass the seeds intact in their fecal matter, which can act as a fertilizer. Dry fruits are usually adapted for wind dispersal of seeds, as for example with the assistance of winglike structures or a fluffy pappus that provides buoyancy. The diversity of fruiting bodies reflects in part the diversity of dispersal agents in the environment, which select for different fruit size, shape, and chemistry.

Pollination and Pollinators

Pollination is the movement of pollen from the stamens to the stigma, where germination and growth of the pollen tube occur. Most (approximately 96 percent) of all flowering plant species are hermaphroditic (possess both sexual functions within a plant, usually within every flower), and thus an individual can be pollinated by its own pollen or by pollen from another individual. Seed produced through self-pollination ("selfed" seed) is often inferior in growth, survival, and fecundity to seed produced through outcross pollination ("outcrossed" seed). As a result, in most species there is strong natural selection to maximize the proportion of outcrossed seed (the "outcrossing rate").

Flowering plants are unusual among seed plants in their superlative exploitation of animals (primarily insects) as agents of outcross pollination. The outcross pollination efficiency of insects, birds, and mammals (primarily bats) may have contributed to both the abundance and diversity of flowering plants. Abundance may have increased because of less wastage of energy and resources on unsuccessful pollen and ovules. Diversity may have increased for two reasons. First, insects undoubtedly have selected for a wide variety of floral forms that provide different rewards (pollen and nectar) and are attractive in appearance (colour juxtaposition, size, shape) and scent (sweet, skunky) in different ways to different pollinators. Second, faithfulness of pollinators to particular familiar flowers may have reduced hybridization and speeded evolutionary divergence and the production of new species.

Although flowering plants first appeared after most of the major groups of insects had already evolved, flowering plants probably caused the evolution of many new species within these groups. Some new insect groups, such as bees and butterflies, originated after flowering plants, their members developing mouthpart structures and behaviour specialized for pollination. In extreme cases, a plant is completely dependent on one insect species for pollination, and the insect is completely dependent on one plant species for food. Such tight interdependency occurs rarely but is well documented in yuccas/yucca moths, senita cacti/senita moths. In all three insects, females lay eggs in the flowers, and their young hatch later to feed on the mature fruit and its contents. Females ensure that the fruit develops by gathering pollen from another plant and transporting it to the stigma of the flower holding their eggs. Plants benefit greatly in outcrossed seed produced, at the small cost of some consumed fruit and seeds, and the insects benefit greatly from the food supply for developing larvae at the small cost of transporting pollen the short distances between plants.

Pollinating agents, whether biotic or abiotic, have exerted strong selection on all aspects of the flower, resulting in the evolution of tremendous floral diversity. This diversity has been distilled into a small number of characteristic pollination syndromes.

Pollination by beetles selects usually for white colour, a strong fruity scent, and a shallow, bowl-shaped flower. Bees select for yellow or blue/purple colorings, a landing platform with colour patterns that guide the bee to nectar (often reflecting in the ultraviolet range of the spectrum), bilateral symmetry, and a sweet scent. Butterflies select for many colors other than yellow, a corolla (petal) tube with nectar at the base, and the absence of any scent. Moths in contrast select for nocturnally opening flowers with a strong scent and drab or white colour, and also a tube with nectar at the base. Bats select also for nocturnally opening flowers, but with a strong musky scent and copious nectar, positioned well outside the foliage for easy access, and drab or white colour. Hummingbirds select for red or orange flowers with no scent, copious nectar production, and a corolla tube with nectar at the base. Other pollinating birds that do not hover while feeding select for strong perches and flowers capable of containing copious nectar (tubes, funnels, cup shapes).

Wind as a pollinating agent selects for lack of colour, scent, and nectar; small corolla; a large stigmatic surface area (usually feathery); abundantly produced, buoyant pollen; and usually erect styles and limp, hanging stamens. In addition there is great floral diversity within any of these syndromes, arising from the diverse evolutionary histories of the member plant species.

Selfing and Outcrossing

Most flowering plant species reproduce primarily by outcrossing, including the great majority of trees, shrubs, and perennial herbs. Adaptations that prevent self-fertilization include self-incompatibility (genetic recognition and blocking of self-pollen) and dioecy (separate male and female individuals). Adaptations that reduce the chances of self-pollination in hermaphrodites include separation of the anthers and stigma in space (herkogamy) or time (dichogamy). In many species, both self-incompatibility and spatiotemporal separation of the sex organs occur.

The ability to produce seeds by selfing, however, is advantageous in situations where outcrossing pollination is difficult or impossible. These include harsh environments where pollinators are rare or unpredictable, and regularly disturbed ground where survivors often end up isolated from each other. Selfing is also cheaper than outcrossing, because selfers can become pollinated without assistance from animals and therefore need not produce large, attractive flowers with abundant nectar and pollen rewards.

Most primarily selfing species are small annuals in variable or disturbed habitats, with small, drab flowers. Most desert annuals and roadside weeds, for example, are selfers. The evolutionary transition from outcrossing to near-complete selfing has occurred many times in flowering plants. Outcrossing and selfing species differ in their evolutionary potential. Outcrossers are generally more genetically diverse and produce lineages that persist over long periods of evolutionary time, during which many new species are formed. Selfers, however, are less genetically diverse and tend to accumulate harmful mutations. They typically go extinct before they have an opportunity to evolve new species.

Asexual Reproduction

The ability to produce new individuals asexually is common in plants. Under appropriate experimental conditions, nearly every cell of a flowering plant is capable of regenerating the entire plant. In nature, new plants may be regenerated from leaves, stems, or roots that receive an appropriate stimulus and become separated from the parent plant. In most cases, these new plants arise from undifferentiated parenchyma cells, which develop into buds that produce roots and shoots before or after separating from the parent.

New plants can be produced from aboveground or belowground horizontal runners (stolons of strawberries, rhizomes of many grasses), tubers (potato, Jerusalem artichoke, dahlia), bulbs (onion, garlic), corms (crocus, gladiola), bulbils on the shoot (lily, many grasses), parenchyma cells in the leaves (Kalanchoe, African violet, jade plant) and inflorescence (arrowhead). Vegetative propagation is an economically important means of replicating valuable agricultural plants, through cuttings, layering, and grafting. Vegetative reproduction is especially common in aquatic vascular plants (for example, surfgrass and eelgrass), from which fragments can break off, disperse in the current, and develop into new whole plants.

A minority of flowering plants can produce seeds without the fusion of egg and sperm (known as parthenocarpy or agamospermy). This occurs when meiosis in the ovule is interrupted, and a diploid egg cell is produced, which functions as a zygote without fertilization. Familiar examples include citrus, dandelion, hawkweed, buttercup, blackberry/raspberry, and sorbus. Agamospermous species are more common at high elevations and at high latitudes, and nearly all have experienced a doubling of their chromosome number (tetraploidy) in their recent evolutionary history. These species experience evolutionary advantages and disadvantages similar to those of selfers.

Evolutionary Significance of Plant Reproduction Strategies

The attractive, colorful, and unique features of the most abundant and diverse group of land plants—the flowering plants—are believed to have evolved primarily to maximize the efficiency and speed of outcross reproduction. Each major burst of angiosperm evolution was a coevolutionary episode with associated animals, primarily insects, which were exploited to disperse pollen and seeds in ever more efficient and diverse ways.

The first major burst of flowering plant evolution was the appearance of the closed carpel together with showy flowers that were radially symmetrical. The closed carpel prevented self-fertilization through recognition and blocking of self pollen within the specialized conducting tissue of the style. Insects attracted to the showy flowers carried pollen between plants less wastefully than wind, and the radial symmetry accommodated insects of many sizes and shapes.

The second major burst was the appearance of bilaterally symmetrical flowers, which happened independently in many groups of plants at the same time that bees evolved. Bilateral symmetry forced bees to enter and exit flowers more precisely, promoting even more efficient outcross pollen transfer. The third major burst of flowering plant evolution was the appearance of nutritious, fleshy fruits and seeds, coincident with a diversification of birds and rodents. The exploitation of vertebrates for fruit and seed dispersal resulted in less haphazard transport of offspring to neighboring populations of the same species (also visited as a food source), thereby reducing the chances that progeny inbreed with their siblings and parents and providing more assurance than wind currents that they find good habitat and unrelated mating partners of the same species.

THE SEXUAL PLANT LIFE CYCLE

Sexual reproduction is important in providing genetic variability. All plants that reproduce sexually must go through meiosis. Meiosis is a unique kind of cell division during which the paired sets of chromosomes present in sexually mature plants, called sporophytes, are halved. Because they have the two sets of chromosomes, one from the male parent and the other from the female, the cells of sporophytes are called *diploid*.

During this process, pairs of homologous, or identical, chromosomes, one from each parent, line up together. Crossing-over, or the exchange of genetic material between these homologous chromosomes, may occur at this time. Crossing-over is critical for producing some of the genetic variability in resulting offspring.

Meiosis typically produces four haploid cells, each with one set of chromosomes, from a single diploid cell. These haploid cells become gametes-eggs and sperm. Without meiosis, sexually reproducing organisms would not have a mechanism for reducing the total chromosome number by half so that genetic variability can be introduced via crossing-over. Meiosis is a critical step that must occur prior to the fusion of sperm and egg, which restores the diploid chromosome number.

ALTERNATING GENERATIONS

Plants undergo a two-phase cycle of sexual reproduction known as the *alternation of generation*. This sexual life cycle involves alternation of the diploid sporophyte generation with the haploid gametophyte generation. The sporophyte stage begins with fusion of an egg and sperm, which produces a zygote. The diploid zygote develops into a sporophytic plant. This sexually mature plant eventually produces meiocytes, cells that will undergo meiosis, typically resulting in four haploid cells. These haploid cells, in turn, develop into gametophytes, which eventually form sex organs where eggs and sperm are produced. The sporophyte and gametophyte stages usually are easy to tell apart; one is typically parasitic or dependent upon the other. Which generation is dominant differs in different plants. Generally, the gametophytes of primitive land plants such as mosses are dominant and larger than the sporophytes. In higher, or more advanced plants, the reverse is true. In other words, the alternation of generations is the alternation between meiocytes undergoing meiosis and the fusion of gametes that produces a zygote.

The Gametophyte Stage

The gametophyte generation normally begins with a spore (sexual spore or meiospore) and ends with a gamete. The cells of the gametophyte are usually haploid.

The Sporophyte Stage

The sporophyte generation usually begins with a zygote and ends with a meiocyte. The cells of the sporophyte are usually diploid.

Reproduction in Fern

Sporophyte development is very rapid in common garden and woodland ferns. The zygote develops an embryonic root, leaf, stem, and foot. The foot apparently aids in absorbing nutrients from the gametophyte. The stem develops slowly, but ultimately produces leaves. The first few leaves of juvenile ferns differ from those of the mature sporophyte. As growth continues, the leaves finally begin to resemble those of the adult species.

Life Cycle of a Fern Diagram.

When the minute sporophyte has become established, the gametophyte dies. The gametophyte generation in ferns, as in other seedless vascular plants, is relatively short-lived and simple. The sporophytic phase is dominant in the life cycle. As in all vascular plants, what we recognize as "the plant" is the sporophytic generation, and at maturity it develops sporangia, the spore-producing organs.

In the common ferns, sporangia are borne on the undersides of fronds, or on modified fronds. The first fronds produced in spring are sterile-that is, they lack reproductive structures. The fertile fronds arise later in the season. In some species, such as lady fern (*Athyrium filix-femina*) the sterile and fertile fronds are similar. In other species-sensitive fern (*Onoclea sensibilis*), for example-the sterile and fertile fronds are very different. The fertile regions on fronds that bear sporangia are known as *receptacles,* and the group of sporangia on a single receptacle is called a sorus. In many species, but by no means in all, the sorus is covered during development by a flap of tissue called an *indusium*. The shape and arrangement of the sori are different in different species.

Spores arise by meiosis from sporocytes. For the most part, immature sori are a pale, whitish colour, although there are exceptions, such as autumn fern (*Dryopteris erythrosora*), whose sori, which are produced in the fall, are bright red. As the spores inside each sporangium mature, they typically get darker, until they are a deep brown or black. Not all spores mature to a deep brown or black, however. Polypodiums are ripe when they are buttercup yellow; osmundas, when they are green. The sporangia of ferns is typically a thin-walled case, usually on a stalk, and surrounded by a ring of thick-walled cells known as the *annulus*, which aids in opening the sporangium when the spores are mature.

The indusium, if present, ultimately shrivels when the spores have matured, exposing the sporangia. The annulus breaks near the base of one side, tearing the sporangium apart, and arches backward. Then the annulus snaps forward abruptly, catapulting the spores into the air. Ferns produce tremendous numbers of spores, but the special requirements of most species for moisture and shade effectively reduce the number of gametophytes that ultimately develop. Spores that are deposited by air currents on sufficiently moist soil and rocks germinate within five to six days, and the gametophyte begins to develop. Most fern spores also require light for germination; those of the bracken fern (*Pteridium aquilinum*) are an exception. Early in germination, colourless rhizoidal cells form at the base of the filament of green cells. These rhizoids, or root-like hairs, absorb and carry water and nutrients to the developing gametophyte. Eventually, the gametophyte becomes heart-shaped, and in some species may be as big as half an inch in diameter. The fern gametophyte was long ago called the *prothallus* or *prothallium*, because it was known to be the precursor of the fern plant, even before its sexual function was clearly understood.

As well-nourished, bisexual gametophytes mature they develop antheridia and archegonia, normally on their undersides. The female, egg-producing organs, or archegonia, occur near the notch of the heart-shaped prothallus. The male, sperm producing organs, or antheridia, occur on the "wings" and opposite the notch. Each archegonium has a chimney-like protuberance that is flared at the top to receive the sperm. Each antheridium is a minute, capsule-like sac where the sperm grow. The cells inside the neck of a mature archegonium dissolve, creating a moist passageway through which the sperm can swim to the mature eggs. The eggs of several archegonia may be fertilized, but usually only one of the zygotes develops into a juvenile sporophyte.

It is interesting to note that while bisexual prothalli may be capable of fertilizing themselves, experiments have shown that the resulting sporophytes often fail to thrive. This suggests that cross-fertilization, the union of sperm and egg from different prothalli, necessary for increasing genetic variability within species, is promoted in at least some ferns.

SEXUAL REPRODUCTION

Sexual reproduction is characterized by processes that pass a combination of genetic material to offspring, resulting in increased genetic diversity. The main two processes are: meiosis, involving the halving of the number of chromosomes; and fertilization, involving the fusion of two gametes and the restoration of the original number of chromosomes. During meiosis, the chromosomes of each pair usually cross over to achieve homologous recombination.

The evolution of sexual reproduction is a major puzzle. The first fossilized evidence of sexually reproducing organisms is from eukaryotes of the Stenian period, about 1 to 1.2 billion years ago. Sexual reproduction is the primary method of reproduction for the vast majority of macroscopic organisms, including almost all animals and plants. Bacterial conjugation, the transfer of DNA between two bacteria, is often mistakenly confused with sexual reproduction, because the mechanics are similar.

A major question is why sexual reproduction persists when parthenogenesis appears in some ways to be a superior form of reproduction. Contemporary evolutionary thought proposes some explanations. It may be due to selection pressure on the clade itself—the ability for a population to radiate more rapidly in response to a changing environment through sexual recombination than parthenogenesis allows. Alternatively, sexual reproduction may allow for the "ratcheting" of evolutionary speed as one clade competes with another for a limited resource.

In the first stage of sexual reproduction, "meiosis," the number of chromosomes is reduced from a diploid number (2n) to a haploid number (n). During "fertilization," haploid gametes come together to form a diploid zygote and the original number of chromosomes (2n) is restored.

PLANTS

Animals typically produce male gametes called sperm, and female gametes called eggs and ova, following immediately after meiosis. With the gametes produced directly by meiosis. Plants on the other hand have mitosis occurring in spores, which are produced by meiosis. The spores germinate into the gametophyte phase. The gametophytes of different groups of plants vary in size; angiosperms have as few as three cells in pollen, and mosses and other so called primitive plants may have several million cells. Plants have an alternation of generations where the sporophyte phase is succeeded by the gametophyte phase. The sporophyte phase produces spores within the sporangium by meiosis.

FLOWERING PLANTS

Flowering plants are the dominant plant form on land and they reproduce by sexual and asexual means. Often their most distinguishing feature is their reproductive organs, commonly called flowers. The anther produces male gametophytes, the sperm is produced in pollen grains, which attach to the stigma on top of a carpel, in which the female gametophytes (inside ovules) are located. After the pollen tube grows through the carpel's style, the sex cell nuclei from the pollen grain migrate into the ovule to fertilize the egg cell and endosperm nuclei within the female gametophyte in a process termed double fertilization. The resulting zygote develops into an embryo, while the triploid endosperm (one sperm cell plus two female cells) and female tissues of the ovule give rise to the surrounding tissues in the developing seed. The ovary, which produced the female gametophyte(s), then grows into a fruit, which surrounds the seed(s). Plants may either self-pollinate or cross-pollinate. Nonflowering plants like ferns, moss and liverworts use other means of sexual reproduction.

Ferns

Ferns typically produce large diploid sporophytes with rhizomes, roots and leaves; and on fertile leaves called sporangium, spores are produced. The spores are released and germinate to produce short, thin gametophytes that are typically heart shaped, small and green in colour. The gametophytes or thallus, produce both motile sperm in the antheridia and egg cells in separate archegonia. After rains or when dew deposits a film of water, the motile sperm are splashed away from the antheridia, which are normally produced on the top side of the thallus, and swim in the film of water to the antheridia where they fertilize the egg. To promote out crossing or cross fertilization the sperm are released before the eggs are receptive of the sperm, making it more likely that the sperm will fertilize the eggs of different thallus. A zygote is formed after fertilization, which grows into a new sporophytic plant. The condition of having separate sporephyte and gametophyte plants is called alternation of generations. Other plants with similar reproductive means include the *Psilotum, Lycopodium, Selaginella* and *Equisetum*.

Bryophytes

The bryophytes, which include liverworts, hornworts and mosses, reproduce both sexually and vegetatively. They are small plants found growing in moist locations and like ferns, have motile sperm with flagella and need water to facilitate sexual reproduction. These plants start as a haploid spore that grows into the dominate form, which is a multicellular haploid body with leaf-like structures that photosynthesize. Haploid gametes are produced in antherida and archegonia by mitosis. The sperm released from the antherida respond to chemicals released by ripe archegonia and swim to them in a film of water and fertilize the egg cells thus producing a zygote. The zygote divides by mitotic division and grows into a sporophyte that is diploid. The multicellular diploid sporophyte produces structures called spore capsules, which are connected by seta to the archegonia. The spore capsules produce spores by meiosis, when ripe the capsules burst open and the spores are released. Bryophytes show considerable variation in their breeding structures and the above is a basic outline. Also in some species each plant is one sex while other species produce both sexes on the same plant.

Fungi

Fungi are classified by the methods of sexual reproduction they employ. The outcome of sexual reproduction most often is the production of resting spores that are used to survive inclement times and to spread. There are typically three phases in the sexual reproduction of fungi: plasmogamy, karyogamy and meiosis.

Insects

Insect species make up more than two-thirds of all extant animal species, and most insect species use sex for reproduction, though some species are facultatively parthenogenetic. Many species have sexual dimorphism, while in others the sexes look nearly identical. Typically they have two sexes with males producing spermatozoa and females ova. The ova develop into eggs that have a covering called the chorion, which forms before internal fertilization. Insects have very diverse mating and reproductive strategies most often resulting in the male depositing spermatophore within the female, which stores the sperm until she is ready for egg fertilization. After fertilization, and the formation of a zygote, and varying degrees of development; the eggs are deposited outside the female in many species, or in some, they develop further within the female and live born offspring are produced.

Mammals

There are three extant kinds of mammals: Monotremes, Placentals and Marsupials, all with internal fertilisation. In placental mammals, offspring are born as juveniles: complete animals with the sex organs present although not reproductively functional. After several months or years, the sex organs develop further to maturity and the animal becomes sexually mature. Most female mammals are only fertile during certain periods during their estrous cycle, at which point they are ready to mate. Individual male and female mammals meet and carry out copulation. For most mammals, males and females exchange sexual partners throughout their adult lives.

Male

The male reproductive system contains two main divisions: the penis, and the testicles, the latter of which is where sperm are produced. In humans, both of these organs are outside the abdominal cavity, but they can be primarily housed within the abdomen in other animals (for instance, in dogs, the penis is internal except when mating). Having the testicles outside the abdomen best facilitates temperature regulation of the sperm, which require specific temperatures to survive. Sperm are the smaller of the two gametes and are generally very short-lived, requiring males to produce them continuously from the time of sexual maturity until death. Prior to ejaculation the produced sperm are stored in the seminal vesicle, a small gland that is located just behind the bladder.

Female

The female reproductive system likewise contains two main divisions: the vagina and uterus, which act as the receptacle for the sperm, and the ovaries, which produce the female's ova. All of these parts are always internal. The vagina is attached to the uterus through the cervix, while the uterus is attached to the ovaries via the Fallopian tubes. At certain intervals, the ovaries release an ovum, which passes through the fallopian tube into the uterus.

If, in this transit, it meets with sperm, the sperm penetrate and merge with the egg, fertilizing it. The fertilization usually occurs in the oviducts, but can happen in the uterus itself. The zygote then implants itself in the wall of the uterus, where it begins the processes of embryogenesis and morphogenesis. When developed enough to survive outside the womb, the cervix dilates and contractions of the uterus propel the fetus through the birth canal, which is the vagina.

The ova, which are the female sex cells, are much larger than the sperm and are normally formed within the ovaries of the fetus before its birth. They are mostly fixed in location within the ovary until their transit to the uterus, and contain nutrients for the later zygote and embryo. Over a regular interval, in response to hormonal signals, a process of oogenesis matures one ovum which is released and sent down the Fallopian tube. If not fertilized, this egg is flushed out of the system through menstruation in humans and other great apes and reabsorbed in other mammals in the estrus cycle.

Gestation

Gestation, called *pregnancy* in humans, is the period of time during which the fetus develops, dividing via mitosis inside the female. During this time, the fetus receives all of its nutrition and oxygenated blood from the female, filtered through the placenta, which is attached to the fetus' abdomen via an umbilical cord. This drain of nutrients can be quite taxing on the female, who is required to ingest slightly higher levels of calories. In addition, certain vitamins and other nutrients are required in greater quantities than normal, often creating abnormal eating habits. The length of gestation, called the gestation period, varies greatly from species to species; it is 40 weeks in humans, 56–60 in giraffes and 16 days in hamsters.

Birth

Once the fetus is sufficiently developed, chemical signals start the process of birth, which begins with contractions of the uterus and the dilation of the cervix. The fetus then descends to the cervix, where it is pushed out into the vagina, and eventually out of the female. The newborn, which is called an infant in humans, should typically begin respiration on its own shortly after birth. Not long after, the placenta is passed as well. Most mammals eat this, as it is a good source of protein and other vital nutrients needed for caring for the young. The end of the umbilical cord attached to the young's abdomen eventually falls off on its own.

Monotremes

Monotremes, only five species of which exist, all from Australia and New Guinea, lay eggs. They have one opening for excretion and reproduction called the cloaca. They hold the eggs internally for several weeks, providing nutrients, and then lay them and cover them like birds. After less than two weeks the young hatches and crawls into its mother's pouch, much like marsupials, where it nurses for several weeks as it grows.

Marsupials

Marsupials reproductive systems differ markedly from those of placental mammals. Females have two vaginas, both of which open externally through one orifice but lead to different compartments within the uterus. Males generally have a two-pronged penis, which corresponds to the females' two vaginae. The penis is used only for discharging semen into females, and is separate from the urinary tract. Both sexes possess a cloaca, which is connected to a urogenital sac used to store waste before expulsion. The female develops a kind of yolk sack in her womb which delivers nutrients to the embryo. Embryos of bandicoots, koalas and wombats additionally form placenta-like organs that connect them to the uterine wall, although the placenta-like organs are smaller than in placental mammals and it is not certain that they transfer nutrients from the mother to the embryo.

Pregnancy is very short, typically 4 to 5 weeks. The embryo is born at a very young stage of development, and is usually less than 5 cm (2.0 in) long at birth. It has been suggested that the short pregnancy is necessary to reduce the risk that the mother's immune system will attack the embryo.

The newborn marsupial uses its forelimbs (with relatively strong hands) to climb to a nipple, which is usually in a pouch on the mother's belly. The mother feeds the baby by contracting muscles over her mammary glands, as the baby is too weak to suckle. The newborn marsupial's need to use its forelimbs for climbing to the nipple has prevented the forelimbs from evolving into paddles or wings and has therefore prevented the appearance of aquatic or truly flying marsupials (although there are several marsupial gliders).

Fish

The vast majority of fish species lay eggs that are then fertilized by the male, some species lay their eggs on a substrate like a rock or on plants, while others scatter their eggs and the eggs are fertilized as they drift or sink in the water column. Some fish species use internal fertilization and then disperse the developing eggs or give birth to live offspring. Fish that have live-bearing offspring include the Guppy and Mollies or *Poecilia*. Fishes that give birth to live young can be ovoviviparous, where the eggs are fertilized within the female and the eggs simply hatch within the female body, or in seahorses, the male carries the developing young within a pouch, and gives birth to live young.

Fishes can also be viviparous, where the female supplies nourishment to the internally growing offspring. Some fish are hermaphrodites, where a single fish is both male and female and can produce eggs and sperm. In hermaphroditic fish, some are male and female at the same time while in other fish they are serially hermaphroditic; starting as one sex and changing to the other. In at least one hermaphroditic species, self-fertilization occurs when the eggs and sperm are released together. Internal self-fertilization may occur in some other species. One fish species does not reproduce by sexual reproduction but uses sex to produce offspring; *Poecilia formosa* is a unisex species that uses a form of parthenogenesis called gynogenesis, where unfertilized eggs develop into embryos that produce female offspring. *Poecilia formosa* mate with males of other fish species that use internal fertilization, the sperm does not fertilize the eggs but stimulates the growth of the eggs which develops into embryos.

ASEXUAL REPRODUCTION

Asexual propagation can occur in various ways. Many plants produce genetically identical copies of themselves, through a mechanism referred to as asexual reproduction. However, botanists more properly refer to this mechanism as "sexual propagation" or "vegetative propagation" because many hold that only sexual reproduction should be referred to as true reproduction, since this is the only kind of propagation that results in the production of new, genetically unique individuals.

Another type of asexual propagation occurs when plants develop underground stems (or rhizomes) which grow outward, or new shoots which grow upward to form new shoots that are genetically identical to the parent. One such example is the trembling aspen (*Populus tremuloides*), which sometimes develops entire stands of trees growing out of the ground as seemingly individual stems, but are actually genetically identical and interconnected below-ground. Another example is the strawberry (*Fragaria virginiana*), although this species has its vegetative runners above-ground.

Other plants develop bulbils on their stems, which can detach, fall to the ground, and sprout to develop new plants that are genetically identical to the original one. One familiar species that does this is the tiger lily (*Lilium *tigrinum*. Other plants can propagate from twigs or stem pieces that fall from the parent, then lodge into a suitable site and develop into a new plant. The crack willow (*Salix fragilis*) can spread itself along watercourses in this manner (as well as by disseminating seeds). Other non-sexual means of propagation include the production of underground bulbs, corms, and tubers that split into parts, each of which is capable of developing into a new plant. Plants that reproduce in this manner include irises and daffodils. All plant organs have been used for asexual reproduction, but stems are the most common.

STEMS

In some species, stems arch over and take root at their tips, forming new plants. The horizontal above-ground stems (called stolons) of the strawberry (shown here) produce new daughter plants at alternate nodes.

Underground Stems

- Rhizomes
- Bulbs
- Corms and
- Tubers

Irises and day lilies, for example, spread rapidly by the growth of their rhizomes.

Leaves

This photo shows the leaves of the common ornamental plant Bryophyllum (also called Kalanchoë). Mitosis at meristems along the leaf margins produce tiny plantlets that fall off and can take up an independent existence.

Roots

Some plants use their roots for asexual reproduction. The dandelion is a common example. Trees, such as the poplar or aspen, send up new stems from their roots. In time, an entire grove of trees may form—all part of a clone of the original tree.

Plant Propagation

Commercially-important plants are often deliberately propagated by asexual means in order to keep particularly desirable traits (*e.g.*, flower colour, flavour, resistance to disease). Cuttings may be taken from the parent and rooted. Grafting is widely used to propagate a desired variety of shrub or tree.

All apple varieties, for example, are propagated this way. Apple seeds are planted only for the root and stem system that grows from them. After a year's growth, most of the stem is removed and a twig (scion) taken from a mature plant of the desired variety is inserted in a notch in the cut stump (the stock). So long the cambiums of scion and stock are united and precautions are taken to prevent infection and drying out, the scion will grow. It will get all its water and minerals from the root system of the stock. However, the fruit that it will eventually produce with be identical (assuming that it is raised under similar environmental conditions) to the fruit of the tree from which the scion was taken.

Apomixis

Citrus trees and many other species of angiosperms use their seeds as a method of asexual reproduction; a process called apomixis.

- In one form, the egg is formed with 2n chromosomes and develops without ever being fertilized.
- In another version, the cells of the ovule (2n) develop into an embryo instead of—or in addition to—the fertilized egg.

Hybridization between different species often yields infertile offspring. But in plants, this does not necessarily doom the offspring. Many such hybrids use apomixis to propagate themselves. The many races of Kentucky bluegrass growing in lawns across North America and the many races of blackberries are two examples of sterile hybrids that propagate successfully by apomixis.

- The female cones of their own species (rare) or
- Those of a much more common species of cypress.

Breeding apomictic crop plants

Many valuable crop plants (*e.g.*, corn) cannot be propagated by asexual methods like grafting. Agricultural scientists would dearly love to convert these plants to apomixis: making embryos that are genetic clones of themselves rather than the product of sexual reproduction with its inevitable gene reshuffling. After 20 years of work, an apomictic corn (maize) has been produced, but it does not yet produce enough viable kernels to be useful commercially.

FACTORS AFFECTING PLANT REPRODUCTION

Climate, soil moisture, soil nutrients, disturbances (such as defoliation, fire, cultivation, and disease) are some of the more common things that affect the reproductive ability of plants. Tolerance versus Requirements: not all plants respond the same to stress. For example, both Agrostis tenuis and Festuca ovina have the same tolerance limits for soil pH, but in a stand of Agrostis and Festuca, Agrostis had a maximum abundance at pH 4-6 and Festuca had a maximum abundance at pH 3-4 and 6.5-7.5. There is also a difference within species. For example, 2 ecotypes of Dactylis glomerata (Norwegian and Portugese) have different temperature requirements for reproduction. In Portugal, it is a winter-grown crop (lower temps) and in Norway it is a summer crop (higher temps) and they respond differently when grown in each other's typical environment.

Plant Community Development

Plant Community - "a group of plants living under relatively similar conditions in a definite area"

Development-"to expand or bring out the potentialities; to advance from a lower to higher state" If these two are combined, then plant community development could be considered "to expand or bring out the potentialities a group of plants living under relatively similar conditions in a definite area".

Stages of Community Development

There are numerous papers on the stages of community development: Succession. Nudation, migration, ecesis (adjustment of a plant to a new habitat), competition, reaction and stabilization.

Disturbance

When discussing community development, it implies that there has been some sort of disruption in the vegetation of the community. Grime defined disturbance as "the mechanisms, which limit the plant biomass by causing its partial or total destruction". Disturbance should always be specified according to spatial extent, temporal extent and magnitude.

Type of Disturbance

Grubb summarized information for different types of plants under different disturbances. While he thought it possible to generalize about type of disturbance and plant life form it was difficult to generalize life characteristics and high disturbance levels. For example, demographic characteristics of communities exposed to grazing reveal a high number of perennial plants capable of both sexual and vegetative reproduction. Annuals and biennials are generally located in interstices and fluctuate in relative abundance.

Grazing intensity tends to increase the dominance of clonal plants. The number of species with vegetative reproduction were significantly different under grazing than those with vegetative reproduction. In the same study, seed dispersal was also analysed. Mobile and adhesion, as methods of dispersal, were both significantly different under grazing. Wind dispersal was significantly different under soil disturbance.

On a denuded glacial foreland, pioneer species were plants whose dominant reproductive form was vegetative. Unlike early colonizing species of secondary successions, pioneers of primary succession are usually long-lived perennials that are relatively slow growing.

Research by Vose and White in a burned Ponderosa pine (Pinus ponderosa) community indicated that buried seed did not contribute significantly to seedling recruitment, and seed rain was more important than buried seed. Most regeneration was from seed rain, and the only vegetative reproduction was from the shrub Ceanothus fendleri. It should be noted that this shrub also reproduced by seed. Allogenic disturbances reveal a question to consider: can we take research from autogenic disturbances and extrapolate them to allogenic disturbances?

Intensity and Frequency of Disturbance

Key characteristics suited to intensity of disturbance are very different. Grubb thought special attention should be placed on plants that were able to increase their range after a disturbance and more study was required on life form, phenolgy, and requirements and tolerances.

Vose and White found that intense heat from a pole-stand burn reduced seed rain compared to sites that burned with less intensity. Burning prevented germination of any buried seed. In their study, the authors speculated that post-fire environmental conditions might have been unfavourable to germination and establishment. It is commonly thought that ruderals, with high seed production, generally dominate areas of frequent, high disturbance. Seeds are usually dormant, and these plants are largely opportunistic of the increases in light supply and nutrients in a disturbed site. A summary of data by Grubb in 1985 supported this theory. In natural areas of yearly disturbance, (for example, scour of drift lines by floods), there is a high number of annuals that use the abundant nutrient supply and set seed in the first season. The seeds of these annuals have no specialized means of dispersal and rely on dispersal though the drift material.

Low Temperatures

On the tundra, there are very low reproductive rates and predominance of vegetative proliferation. Most invading species are either self-pollinated, or invest massive resources in vegetative propagules there is a higher representation of species that possess some mode of uniparental reproduction.

Chapin calculated that the carbon cost of producing a tiller of Dupontia fisheri from seed at Pt. Barrow, Alaska was 10,000 times greater that that of producing tiller vegetatively. A single Betula pubescens in Swedish Lapland produces 40,000,000 seeds thoughout its life, only one of which is necessary to replace the original tree for a stable population. Are these seeds only important in the colonization of new areas outside the forest or after disturbance such as fire within the forest? Recruitment from seed is important in open habitats such as fell-fields, but early mortality rates are high. Vegetative reproduction has low recruitment levels but high survival rates due to the physiological interdependence of modules. (fell fields are open communities of scattered individuals aggregated into islands). The plants are dwarfed and prostrate, and the habitat is characterized by little winter snow cover, water shortage, primitive soils and frequent soil movement due to freeze-thaw cycles).Chambers cautioned in diminishing the importance of sexual reproduction in tundra environments. And though compared to temperate climes, sexual role is reduced, for tundra it is still very important.

Callaghan and Emanuelsson indicate that sexual reproduction may be adapted to the exceptional and not too normal circumstances. Environmental conditions that select for sexual reproduction may be very different from those that select for vegetative growth.

Low nutrients

On the tundra and in alpine disturbances, distinct species colonization patterns can be observed. Chambers noted that species with nitrogen-fixing symbionts are frequent colonizers on mineral soils low in nitrogen, including those exposed by retreating glaciers or snowbanks. In alpine tundra areas of severe soil disturbance, where the resultant soils are poorly developed and have low nutrient status, colonization is most frequently from dissemules, mostly seeds dispersed from adjacent communities. In these conditions, it is necessary for plants to produce many seeds that can easily be dispersed

Importance of Soil and Water

Prach and Pysek studied the role of clonal plants in plots that were 10 to 50 years old, grown under different types of habitats. They had acidic emergent bottoms, peaty soils, sandy, reclaimed, abandoned sand pit, ruderal sites ranging from poor to rich and old fields that were wet to xeric. From their observations of old fields, clonal plants dominated under xeric conditions, while non-clonal dominated under mesic conditions. These sites were 50+ years old. This indicates that more long-term studies in old fields are needed under various environmental conditions.

The study they did for plots that were 10 years old concluded the data set was limited and not able to reveal a functional relationship between the participation of clonal plants and habitats.

Stocklin and Baumler found that seeds found in areas around a glacier (primary to secondary successional areas) were extremely dependent on water availability for germination and establishment. Seedling success was dependent on "safe" sites, which were slight indentations or areas covered by dead plant material. Chambers found that in windy, soil exposed areas (tundra/alpine) relationships between soil characteristics and seed attributes significantly affected the spatial distribution of dispersed seeds. This may be one of the primary determinants of species colonization patterns.

Chambers found that with small soil particles, small seeds and seeds with adhesive coats were trapped, but most large seeds moved across the surface;

at large particle size, larger seeds were trapped, with larger particles, small seeds moved further downward in the soil as well. In the same study, surface soil temperatures were measured and on dark covered organic soils, the temperature ranged from 35-45 °C on the surface and 20-25 °C at 5 cm depth. On an adjacent site with exposed, loamy sands, midday soil temperatures were about 5 °C lower. Soil temperature can result in higher rates of mineralization and decomposition. Most tundra species had higher germination rates under higher temperatures than what is normally found in the field.

Factors that Influence Colonization

It has been hypothesized that the role of clonal dispersal in succession is evident both in early stages, due to the rapid capture of space, and in late-successional stages, due to the rapid filling of gaps in more or less closed vegetation cover. In essentially closed cover, seeds do not have the opportunity to germinate, nor do seedlings have the chance to establish. Competition for nutrients, water and light is just too great. This would imply that eventually all plant communities should be dominated by plants that rely heavily on vegetative reproduction. Why is it then, that there are many plant communities that appear to stabilize with a combination of plants that use both seed (sexual) and vegetative (asexual) reproduction? Research on the role of clonal species is lacking, and an understanding of their behaviour is based to a large extent on speculation. The factors that inhibit or promote one reproductive strategy over another are not well understood.

Index